FAREWELL FOSSIL FUELS

Reviewing America's Energy Policy

FAREWELL FOSSIL FUELS

Reviewing America's Energy Policy

SIDNEY BOROWITZ

PLENUM TRADE • NEW YORK AND LONDON

Library of Congress Cataloging in Publication Data

Borowitz, Sidney, 1918–
Farewell fossil fuels: reviewing America's energy policy / Sidney Borowitz.
p. cm.
Includes bibliographical references and index.
ISBN 0-306-45780-6 (hardbound).—ISBN 0-306-45781-4 (pbk.)
1. Renewable energy sources—United States. 2. Energy policy—United States. 3. Fossil fuel—United States. I. Title.
TJ807.9.U6B67 1998
333.79'4'0973—DC21

98-5671
CIP

ISBN 0-306-45780-6 (Hardbound)
ISBN 0-306-45781-4 (Paperback)

A Division of Plenum Publishing Corporation
233 Spring Street, New York, N.Y. 10013-1578

http://www.plenum.com

10 9 8 7 6 5 4 3 2 1

Printed in the United States of America

CONTENTS

PREFACE

This is a book about the dangers of having fossil fuels as the principal energy source of the world. It did not start out that way. At first, it was my intention to use a collaborator and write a book about energy for the general public but make it also suitable as a text or auxiliary text for a course on science for nonscience students. In the course of the writing I felt strongly about changing the focus of the book urging the revision of the energy policy of the United States, and indeed almost all of the countries of the world. I felt that such a book would serve a more useful purpose. The result might still be useful as a text for some courses for nonscience majors but it is intended mainly for a larger lay audience who might influence governments to encourage the use of alternative renewal sources of energy as replacements for fossil fuels.

The urgency of the issue of encouraging the development of renewable sources derives from the fact that there is little being done to prepare the public for the consequences of fossil fuels becoming scarce and disappearing altogether. This despite the fact that the global stores of oil are unlikely to last more than a century. Nor is there clear sailing for addressing the more immediate problem of global warming as a result of the greenhouse effect arising from the burning of fossil fuels.

I would like to thank Professor Grace Marmor Spruch of Rutgers University, my initial collaborator on an earlier book on energy, which was not completed. The collaboration clarified many points useful in writing the present volume. I also thank Mr. Douglas Blaufarb for considerable editorial assistance. Finally, thanks are due to Professor Edward Gerjuoy of the University of Pittsburgh for clarifying many technical and scientific details.

Sidney Borowitz

INTRODUCTION

In the 1970s, the Organization of Petroleum Exporting Countries (OPEC), which controlled virtually all of the oil resources of the world outside of the United States, increased the price of crude oil by a factor of three, precipitating a major recession in the United States and indeed in the world. This action gave us some inkling as to what the consequences of an oil crisis could be. Since that time the economies of the world have adjusted to the increased price of oil, and the sense of impending doom has all but disappeared. But the threat of another crisis should still be with us.

The source of more than 90 percent of the energy we use is from fossil fuels, oil, coal, and natural gas. It took billions of years for these supplies to be created. While there is yet no shortage, the amount of resources remaining is still finite. And we have little more than a century to develop adequate, practical, and inexpensive technologies to replace some of these sources. It is time to sound a tocsin to urge us to prepare for the inevitable—that an energy crisis due to scarcity will occur, whose consequences are not as easily overcome as adjusting to increased fuel prices.

Fossil fuels would seem to be a gift from the gods to earthlings. The availability of this source of energy has brought unimaginable comfort and prosperity to many of the inhabitants of our planet. At first glance these fuels seem to be cheap, clean, safe, and almost infinitely available. The case for continuing their use as the engines of our economy seems overwhelming. Indeed, it appears to be the energy policy of the United States to rely on fossil fuels for the foreseeable future. The Clean Air Act of 1993 emphasizes ways of making fossil fuel utilization less burdensome for the environment, and of increasing the efficiency of the devices using fossil fuels. It does not provide for actively supporting projects seeking early replacements for these energy sources.

Upon closer examination, this source is neither as infinite nor as safe, clean, or cheap as one might suppose. The reserves of oil are expected to last only for decades or perhaps as much as a century at the present level of use. The known reserves of natural gas are expected to last equally as long but information about the untapped reserves is not as accurate as is that for oil. Although the reserves of coal are not as short as those of petroleum, still they are a finite resource that, with use, will ultimately disappear. Coal, which is must more plentiful, is a killer. It has taken the lives of about 100,000 miners since the beginning of the century. In addition it is responsible for the many cases of black lung disease among miners. The fine airborne particles of coal dust cause many upper respiratory diseases including lung cancer. Burning coal produces environmental gases that destroy lakes and forests. Obtaining coal by strip mining causes an environmental disaster.

The use of petroleum products in automobiles and trucks is responsible for smog, low-level ozone (an upper respiratory irritant), and excess carbon monoxide in the atmosphere. Oil spills, each of which seems like a preventable accident, are clearly hazards that are part of the oil economy. And they have not been a boon to marine life.

State and federal environmental protection agencies have done a commendable job in dealing with some, but not all, of the environmental challenges presented by fossil fuel use. However, one challenge of fossil fuel use cannot be addressed except by eliminating fossil fuels as an energy source or at least tempering their use. The burning of any fossil fuel results in increasing the carbon dioxide in the atmosphere, an effect that a United Nations scientific advisory committee has stated will lead to global warming, with serious climatic and economic consequences, perils that will be more extensively dealt with in this text. Global warming may have already begun. The future of fossil fuels will be a mixed blessing for the Earth and its inhabitants.

The energy produced by fossil fuels is not as cheap as it seems. The market price does not reflect the political and diplomatic costs inherent in its use. One-fourth of the global reserves of oil are in the Middle East, a warren of satrapies that are difficult to deal with. The flow of oil is so important to the economy of the United States that it is necessary to invest billions to maintain a fighting force should the flow be interrupted. That such an investment is necessary was adequately demonstrated in Operation Desert Storm during the Persian Gulf War against Iraq. And it justifies keeping a reaction force on the ready in that region. Desert Storm was fought because it seemed as if the flow of oil to the United States and worldwide could be seriously interrupted. Imagine the danger to the peace on Earth when the supplies of petroleum are seriously diminished. Although global warming poses a more immediate threat, the finiteness of the supplies of oil may be even more dangerous in the long run. More than half of the oil used in the United States has to be imported. About half of these imports come from the Middle East. They affect the balance of trade of the United States adversely and have serious budgetary consequences. None of these considerations, despite their effect on the pocketbooks of the citizens, is included in calculating the cost of using fossil fuels. In 1998 the reaction force was strengthened in response to Iraq's refusal to allow unfettered inspections of potential sites that might be loci for the manufacture of nuclear or biological weapons of mass destruction.

Although public policy has focused on maintaining fossil fuels as a source of energy, there has been some support for the development of alternatives. The cost of generating energy commercially using these alternatives is rapidly decreasing. Still, no matter how cheap the alternatives, they would have great difficulty penetrating the energy market to any appreciable extent. For one thing, the market price of fossil fuels is unreasonably cheap. Their cost for the most part reflects only the cost of recovery and delivery, not manufacture. For another reason, the capital investment in the fossil fuel economy has been largely amortized and new technologies would require substantial capital investments to compete. For still another, powerful financial interests are invested in fossil fuels, and they will fight to keep these entrenched as long as possible. The tactics for this fight have already been revealed and will be discussed in the text. For there to be alternatives, there has to be some initial subsidization.

The prevailing public opinion is that the best strategy for dealing with the fossil fuel problem is to encourage conservation. Bowing to this popular point of view this book will discuss the variety of ways in which conservation can lengthen the time for fossil fuels to continue as our preeminent source of energy. With conservation in place for fossil fuels and their use discouraged, a window of opportunity exists while they are inexpensive and plentiful, to begin to replace these carbon-based fuels by other technologies as quickly as possible. With the best of intentions it may take a generation or more to introduce extensive use of alternative energy sources.

Subsidization is especially reasonable because such a step will ultimately have to be taken. The fossil fuels will, without doubt, become scarce first, then expensive, and finally disappear altogether. Estimates as to when this might occur may well be wrong, but as Professor Joel Cohen of the Rockefeller University explains, "the future is unlike the past because it hasn't happened yet." Thus, the sooner we phase out fossil fuel use, the more relaxed we can be about not having to endure the dislocations caused by global warming. The naysayers assert that the case for global warming has not been made. But if the majority of the scientific community is correct, doing nothing now will leave us without an alternative. Prudent policy would dictate providing some help to bring alternative energy sources to market.

It is well to realize that no amount of exhortation or argument is likely to effect a change in our accustomed way of life. Such matters are almost always determined solely by our culture's economic considerations. It is the responsibility of the political leadership to create the economic incentives to effect a change.

In 1993, there was an indication of how conservative we are about change and how important the economic effects of using alternative energy sources can be. This happened when Congress considered an increase in the tax on the use of gasoline. A substantial tax increase would go a long way toward reducing the federal deficit as well as reducing the use of petroleum products, an environmental benefit and a conservation measure. The best the conflicting interests could arrive at in that Congress was a tax increase of 4.3 cents per gallon—much less than the normal fluctuation in gasoline prices during the year. This small tax increase was a hotly debated political issue prior to the 1996 presidential campaign but fortunately a repeal of this increase did not take place.

Our Earth is precious. This book entails how fossil fuel use has damaged Earth and our way of life. We are close to being able to do without fossil fuels by leveling the playing field which will enable the new technologies to compete economically with the existing energy sources. It is hoped this book can influence the public sufficiently for this to occur.

CHAPTER 1

THE EARTH EMERGES

And all men kill the things they love,
By all let this be heard,
Some do it with a bitter look,
Some with a flattering word,
The coward does it with a kiss,
The brave man with a sword!

The Ballad of Reading Gaol,
OSCAR WILDE

Our principal sources of energy are the fossil fuels: coal, oil, and natural gas. We shall in due course describe how fossil fuels became so plentiful and how they got their name. Before we do this, we shall describe how our planet came into being.

We take for granted our habitable Earth, where we live in comfort and prosper. We give little thought to how dicey it was a planet formed and evolved into a fit place for us.

First we had to have a star, the sun, to provide us with exactly the right amount of energy to sustain us. Then we had to be the appropriate distance from this star. The planet had to be the correct mass to hold an atmosphere of just the right size and have the proper composition.

To appreciate how close we came to annihilation, we might compare ourselves with two similar planets, Venus and Mars. Venus is so hot any metal that comes in contact with its atmosphere would melt. Mars is too cold to sustain life as we know it and has no atmosphere. (This is similar to the situation Goldilocks faced in the "Three Bears." When Goldilocks tried the cereal in the three bowls, she found the first was too hot, the second too cold, and the third just right.)

The short answer as to how this happened is that Venus is too close to the Sun, and Mars is too far away (Fig. 1.1, Table 1.1). But that is not the whole story. The real story has to do with the size of these planets and the nature of their primitive atmospheres—whether they were so constituted that it was possible for water to rain down on them, and whether they were large enough to retain the heat generated by their constituents and prevent the planets from freezing.

About 5 billion years ago, there was a cloud of gas and solid particles in the part of the universe now occupied by our solar system. This cloud initially consisted chiefly of hydrogen and helium, the remnants of original creation. These gases do not liquefy or freeze until their temperature is much lower than was the temperature of this particular cloud. The solid particles were frozen pieces of almost all of the elements with which we are familiar and of compounds of these elements. They were solid because the temperature of this cloud was about −220°C (−373°F). The elements had been created by the stars formed in the universe at an earlier time.

At this time the gravitational pull of the particles on each other caused them to begin a general drift toward the center of the cloud. As the particles started to clump together, their temperature went up. This increase in temperature is not unlike a gas becoming warmer as it is compressed. As you might guess, this clump ultimately became our hot sun. The experts do not completely agree on the theories as to how the planets were formed. However, there is no doubt that the planets formed over a span of at least 100 million years by the accretion of particles in the outer regions of the gaseous cloud. During the period of their formation the planets warmed. This happened because: (1) the sun was becoming warmer, (2) radioactive elements within the planets themselves discharged energy into them; and (3) the planets were bombarded by the particles still in the cloud. Soon after its birth, the Earth was hot enough to melt the iron in its

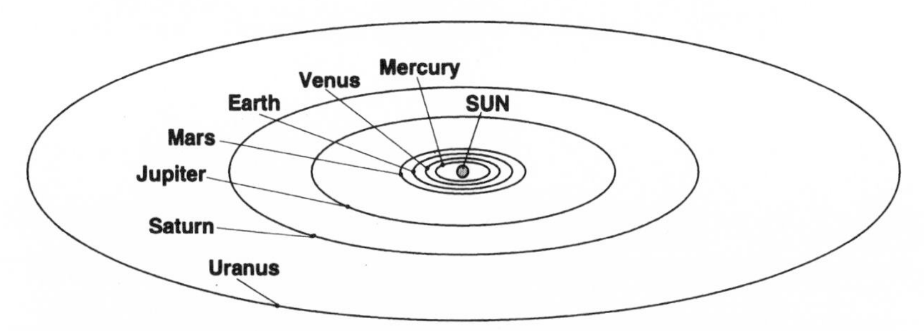

Figure 1.1. Venus, Earth, and Mars are close to each other, compared to their distances from Jupiter. Saturn is twice the distance of Jupiter, and Uranus is much farther still. The actual distances are given in Table 1.1.

surrounding cloud, sinking it to the center, and to melt the silicates which, because they were lighter, formed its crust. After the planets cooled sufficiently there remained an atmosphere consisting principally of carbon dioxide, water vapor, ammonia, methane, and some rare gases such as neon and xenon. The planets closest to the sun (Mercury, Venus, Earth, and Mars) were not massive enough to be able to retain by gravitational attraction the lightest elements, hydrogen and helium, in their atmospheres. The heavier outer planets

Table 1.1. Distances from the Sun

Planet	km × 10^6
Venus	108
Earth	150
Mars	228
Jupiter	778
Saturn	1426
Uranus	2870

Table 1.2. Planet Masses and Diameters

Planet	Mass	Diameter
	kg × 10^{24}	km
Venus	4.9	12,100
Earth	6.0	12,750
Mars	0.64	6800
Jupiter	1900.0	142,800
Saturn	570.0	120,000

(Jupiter, Saturn, Uranus, and Neptune) held on to an atmosphere principally of hydrogen and helium (Table 1.2). Carbon dioxide and water vapor formed when the carbon and hydrogen in the original cloud combined with the very reactive element, oxygen. Oxygen had disappeared into these molecules. With this introduction, we are now ready to address "Goldilocks' choice."

What distinguishes Earth from Venus is that Earth had a large amount of liquid water on its surface in its early stage of development. We have noted that water vapor was one of the components of its atmosphere. But the planet needed liquid water to survive and that was brought into existence in the following manner: During the early history of Earth, the sun's intensity, as estimated by planetary scientists, was 30 percent less than it is now. One can calculate the Earth's temperature with such a sun shining on it—−18°C, below the freezing temperature of water. Water freezes at 0°C.) If any ice were to cover Earth it would be very effective in reflecting the sun's rays, thus causing the Earth's temperature to drop still further. The Earth would become irreversibly ice covered, making it a white Earth catastrophe. Fortunately the carbon dioxide and water vapor in the atmosphere prevented this from happening.

These two gases, the rising intensity of the sun, and the heat produced by the intense internal radioactivity, caused the Earth's temperature to increase to well above water's freezing temperature but below its boiling point. Planetary scientists now believe Earth's temperature rose to at least 15°C (59°F), the present average temperature of Earth. This permitted water vapor in the atmosphere to condense into liquid. The role these gases played in increasing Earth's temperature will be detailed in the next chapters. This role is similar to the one plants and glass play in keeping the inside temperature of a greenhouse higher than the outside temperature of its surroundings. This is called the "greenhouse effect."

The existence of liquid water on the planet facilitated a chemical reaction (weathering) involving the silicate rocks that had become exposed. This reaction removes carbon dioxide from the atmosphere. It is fortunate that it does so. The sun's rays continued to intensify, causing Earth's temperature to rise. However, carbon dioxide was being removed from the atmosphere by weathering, lessening the greenhouse effect. The internal radioactivity was diminishing, offsetting the effects of a more intense sun, but keeping the temperature of the planet steady and within proper bounds to sustain life. What had come about was a perfect self-regulating cycle. There was an adequate supply of water to sustain the weathering because the temperature remained steady. This permitted the water vapor in the atmosphere and the steam coming from Earth's interior to liquify. It is estimated that it might have rained for 100,000 years. This is no surprise considering how much water there is on our planet. Someone once said our planet should be called "Ocean" rather than Earth because its surface consists of more than 70 percent water.

Venus was not so lucky. Although Venus' atmosphere was initially similar to Earth's, it ended up being closer to the sun. As a result, the sun's radiation coupled with Venus' own greenhouse effect, warmed Venus' atmosphere to such high temperatures that water vapor never could condense. The greenhouse effect, due to the water vapor and carbon dioxide in the atmosphere, kept increasing Venus' temperature until it reached the dreadful heights it is at today—900°F. Whatever water might have been emitted from the interior of Venus would have been vaporized as well.

The early history of Mars was quite different. When the surface of Mars was explored, it revealed definite indications that liquid water did condense on its surface. But the liquid water did not last long. Water will remain in the liquid state only if the planet stays warm. The sources of heat for Mars were the greenhouse gases in the atmosphere and the heat supplied by the planet's volcanic action, which also supplied carbon dioxide and its internal radioactivity. Mars, however is a small planet compared to the Earth. It is half its diameter, one-eighth its volume (Table 1.1), and only one-tenth its mass. The smaller the planet, the more rapidly it will lose its greenhouse gases. The less massive the planet, the more rapidly other sources of heat will be dissipated. Thus all avenues for keeping Mars warm were quickly shut down. Mars could not retain its internal heat, and the replenishment process for carbon dioxide disappeared. It then became a frozen world. Once the liquid water turned to ice, the process was irreversible because the ice reflected almost all of the radiation coming from the sun and did not permit the planet to warm up.

There is now some evidence, not universally accepted, that some primitive forms of life might have existed on Mars before the chilling occurred when liquid water was available. If this information is true, it would be consistent with the description we have given of Mars' history.

To summarize, we can see that Earth had to be precisely the right size and distance from the sun for it to develop a habitable environment. It has been estimated that if the Earth were 4.5 million miles closer to the sun, it would suffer the fate of Venus; if it were appreciably smaller, the fate of Mars.

The secret of the development of a habitable atmosphere lies in the availability of liquid water to permit the weathering process, which removes the carbon dioxide in the atmosphere at a rate balancing the warming factor of the sun. As the sun's radiation increased, Earth's temperature was maintained at just the right level to support life. Life has been discovered to have appeared on Earth at least one billion years after it was formed—at least 3.5 billion years ago.

That life existed on Earth 3.5 billion years ago has been known since 1977. Prior to then it was thought life originated much later. Elos Barghoorn, a paleontologist at Harvard University, discovered these early beginnings. He was exploring Fig Leaf, a small South African mountain. He brought back to his university laboratory some pieces of rock he had chipped off from the mountain. Barghoorn then made very thin sections of the rock, thin enough for light to be able to shine through. And behold, in some of the samples he was able to make out the outlines of some fossilized organisms, bacteria or algae. It is possible to date samples of rock fairly accurately. The soil in the vicinity of the rocks he brought back contained radioactive argon, which decays into potassium at a known rate. By measuring the ratio of radioactive argon to potassium in the soil one can accurately determine the age of the rocks. Fig Leaf mountain's rocks turned out to be 3.5 billion years old.

The atmosphere of the early Earth did not have the same composition it has today. The principal difference is that there was no oxygen. There was only carbon dioxide, methane, ammonia, water vapor, and some hydrogen sulfide. The organisms Professor Barghoorn discovered must have been able to subsist by metabolizing these chemicals, probably using the hydrogen as a nutrient. This should come as no surprise to us because even today there are bacteria existing deep in Earth's crust. These are sustained by the sulfides rich in hydrogen coming out of the crevices and cracks between tectonic plates where other forms of life also exist. These forms of life are called anaerobic, meaning they are capable of existing in the absence of oxygen.

It is conjectured that oxygen came into the atmosphere because the resident bacteria created a scarcity of hydrogen sulfide by consuming its hydrogen. When hydrogen sulfide became scarce these bacteria could only get sufficient hydrogen by consuming the ample supply of hydrogen in water. When hydrogen is removed from water, oxygen is released. In this way we get the principal source of the oxygen in our atmosphere today. Lynn Margulis, professor of Biology at the University of Massachusetts at Amherst, has called the freeing of oxygen into the atmosphere "The Oxygen Holocaust," believing it to be the greatest poisonous pollution of the atmosphere in Earth's history. Indeed it must have been for the bacteria living at that time. Oxygen is a corrosive, reactive chemical that could easily destroy an organism whose metabolism is unable to use it. However, some fraction of the resident bacteria must have found a way of detoxifying the oxygen and exploiting it for their metabolism.

Bacteria belong to a class of living things called prokaryotes. These are simple living things that lack cellular nuclei. Higher living things, such as humans, are called eukaryotes. Prokaryotes reproduce by dividing but, unlike us, produce a new generation very quickly, every 20 minutes or so. They also are very adept at creating mutations very quickly by changing their genetic composition. We know this from our present experience with bacteria surviving the antibiotics we throw at them only to find that they soon become resistant. The hydrogen consuming bacteria using water as a source of nutrition continued to supply oxygen to the atmosphere. The more robust oxygen eaters eventually replaced the hydrogen sulfide eaters because their metabolism was more efficient and the supplies of hydrogen sulfide were diminishing.

A number of experiments have been performed indicating that it was possible for life, a form of matter that has the ability to self replicate, to have emerged spontaneously on Earth. Six elements compose 90% of living matter: carbon, oxygen, nitrogen, hydrogen, potassium, and sulfur. The first four were available in large quantities in the primitive atmosphere. These elements are combined into amino acids in living things which, in turn, are combined into proteins, the work-horses of life that make all biological processes happen. These elements are also combined into chemicals called nucleotides which are the building blocks of DNA and RNA. These are the engines of replication which they accomplish by encoding the cipher for the creation of proteins.

In 1953, Stanley Miller, a 23-year-old graduate student of the Nobel Prize winning chemist, Harold Urey, passed an electrical discharge through a soup consisting of ammonia, hydrogen, water vapor, and methane, for a week (Fig. 1.2). This soup was supposed to simulate the atmosphere of primitive Earth. He was rewarded when the soup produced two amino acids and several organic compounds which, until that time, were believed to be producible only by living cells.

Dr. Miller has continued his research as a professor at the University of California at San Diego. In experiments similar, but more sophisticated, than the ones he performed as a graduate student, he has been able to create 13 of the amino acids essential to life. Professor Miller and his collaborators have also demonstrated that the basic nucleotides of RNA could have been formed by processes occurring spontaneously on Earth.

The importance of this discovery is this: of the two substances which encode life's processes, RNA and DNA, the former is simpler, being single-stranded rather than double-stranded. In 1982, Professor Thomas Cech of the University of Colorado discovered that RNA without the assistance of DNA or other enzymes could split up and splice itself to create new arrangements of RNA. This is a process that is an exclusive attribute of living things and was previously thought to be capable of being performed only by DNA. A further experiment suggesting that life could have arisen spontaneously on Earth was reported in the journal *Nature* in May 1994. In this experiment, DNA produced in the laboratory was successfully made to replicate spontaneously in a chemical environment created by the experimenter. This result, and the experiments of Professors Miller and Cech, make it more reasonable to postulate that life could have arisen on earth without any external agency affecting the outcome.

There is always the unsettled question as to whether chance and chance alone could have produced life and, ultimately humans, from the primeval soup originally existing on Earth or whether some guiding hand was needed for this to take place. Most modern molecular biologists' response to this much debated question is that natural selection together with a great deal of time is all that is necessary to account for what we now see. This point of view has been elaborated by Daniel Dennet in his book, *Darwin's Dangerous Idea*. This is the point of view adopted here. But it is fair to say that despite the fact that the Pope announced that Darwin's hypothesis

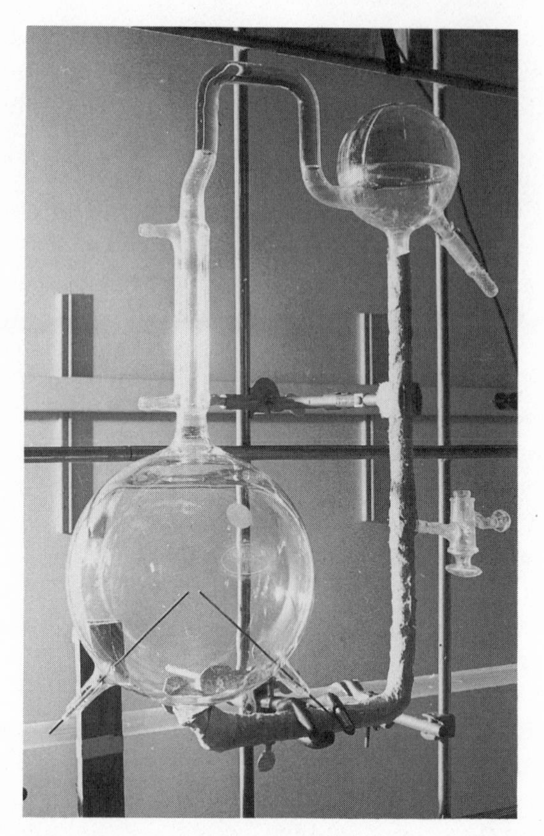

Figure 1.2. Stanley Miller used this simple apparatus to show how complex amino acids could be formed by electrical discharges in an atmosphere resembling the prehistoric one. (Courtesy of Sio Photo.)

is consistent with Catholic doctrine, and that the theory of evolution is taught in Catholic Schools, it is not universally accepted. There are many, perhaps even a majority, who believe the story of creation that appears in religious texts is the correct one. A discussion of the pros and cons of this controversy here would take us too far afield.

So far, we have described the processes that are required to account for only the most primitive forms of life. The more complex forms of life from which the fossil fuels were formed took more than two billion years more to appear on earth.

About 800 million years ago, ancestors of our modern plants appeared on Earth. Dead plant material was ultimately converted into coal by a chemical process which took place over hundreds of millions of years. This metamorphosis also incorporated into coal many substances that have proved harmful to our environment when the coal is burned. Carbon, a ubiquitous component of coal was originally in the plants (growing plants remove carbon dioxide from the atmosphere). It poses a different and more complicated environmental problem and will be discussed in Chapter 3. Oil and natural gas have had a somewhat different history but the consequences of using these as fuel are similar to what they are for coal. Coal, oil, and natural gas, our principal sources of energy, are properly called fossil fuels in view of the way they came about.

In the past few decades a great deal of progress has been made to reverse the environmental damage of the noxious substances released when fossil fuels are burned. Through legislation that encouraged changes in manufacturing processes and the disposal of wastes, our air has become cleaner, our water purer, and our rain less acidic. Although a great deal has been accomplished, much remains to be done. But at least we have confidence that we can make as much progress as our resources can sustain if we only have the will.

A much more difficult problem of environmental pollution is the carbon dioxide emissions that occur when fossil fuels are burned. These are responsible for the greenhouse effect and global warming. The serious threats this poses for the environment will be discussed in Chapter 3.

It is ironic that the long and astounding evolution of the primordial soup into intelligent humans has produced a species that can sufficiently alter its environment into a poisonous one, forcing a drastic and costly change in its way of life. If Oscar Wilde were alive today, could he spare one verse of the "Ballad of Reading Gaol" to describe how humans contribute to the destruction of the thing they should love: the Earth?

If we can go at least part way toward replacing fossil fuels with more benign sources of energy, we can begin to hope that the future will not seem so bleak. This book is intended to point to some paths that can be taken to accomplish that goal without forcing us to make too many painful sacrifices.

CHAPTER 2

ENERGY

There is a great understandable hunger for energy in the world. As Fig. 2.1 illustrates, increased uses of energy are strongly correlated with the gross domestic product, a measure of the average standard of living of the inhabitants of a country. But if the principal sources of energy are fossil fuels, we are faced with a dilemma: burning these fuels contributes to the greenhouse effects and thus to the warming of the Earth, causing serious environmental consequences.

While there are many forms of energy with which we are familiar—mechanical, chemical, nuclear, light, thermal energy, heat, to name a few—and there are many sources for all forms of energy, the inhabitants of our planet have found it convenient to exploit relatively few. The burning of the so-called fossil fuels (coal, oil, and natural gas) currently provide us with about 90 percent of the energy we use. These are finite resources and their use is environmentally harmful. We intend to explore in this book what alternatives are available to us. The purpose of pressing for the ultimate replacement of fossil fuels in our economy is not only to improve our environment but to prepare ourselves for the time when these fuels become scarcer, and therefore more expensive, or have been depleted altogether.

Virtually all of the energy available to us on Earth comes from the sun. (The only exceptions are the relatively small amounts we use from geothermal sources, nuclear energy, and the tides.) Of all of the energy the sun radiates, only a small fraction is directed toward

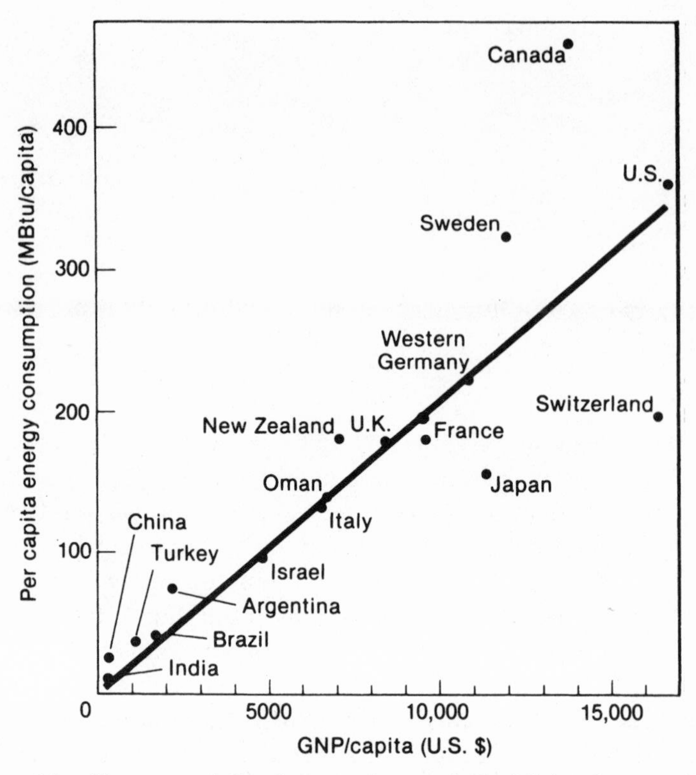

Figure 2.1. Strong correlation between the standard of living in countries as measured by each country's gross domestic product and its per capita energy consumption. (Reprinted with permission from Saunders College Publishing.)

Earth. We are 93 million miles away from the sun and our planet is only 8,000 miles in diameter so we can only intercept a small fraction of the fraction directed toward us (Fig. 2.2). Of the radiation intercepted by Earth, 30 percent is reflected back into space by the atmosphere and never reaches us at all. Another 47 percent is absorbed by the atmosphere either directly or by the greenhouse gases reflecting back to Earth the radiation originally reflected from Earth (Fig. 2.3). This energy is not wasted because it provides us with a warm comfortable climate to live in. Because the atmosphere is not heated

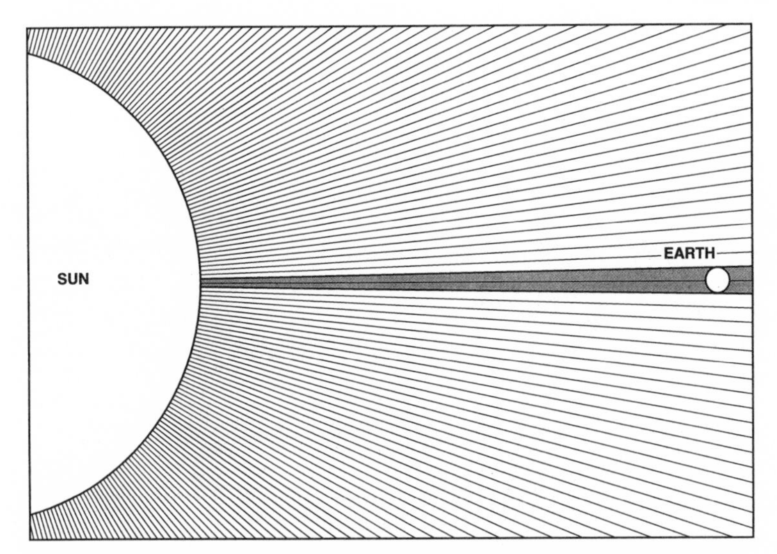

Figure 2.2. Earth is so small and its distance from the Sun is so great, Earth can intercept only a small fraction of the energy the Sun radiates.

uniformly by the energy it absorbs, winds are created by the movement of air from regions of low temperature to regions of high temperature. These winds can be—and are being—used to supply us with energy. A further 23 percent of the energy intercepted goes into the evaporation of water, which ultimately comes down as rain. This, too, is a source of energy on Earth.

In Chapter 1 we described how the greenhouse effect contributed to making the original Earth habitable for us. As we indicated, this same greenhouse effect could potentially interfere with our way of life. Such a conclusion amounts to an indictment of the use of fossil fuels and so deserves a more detailed explanation.

As we have noted, the temperature of the Earth averages 15°C (59°F) and the surrounding outer space is −270°C. A warm body in

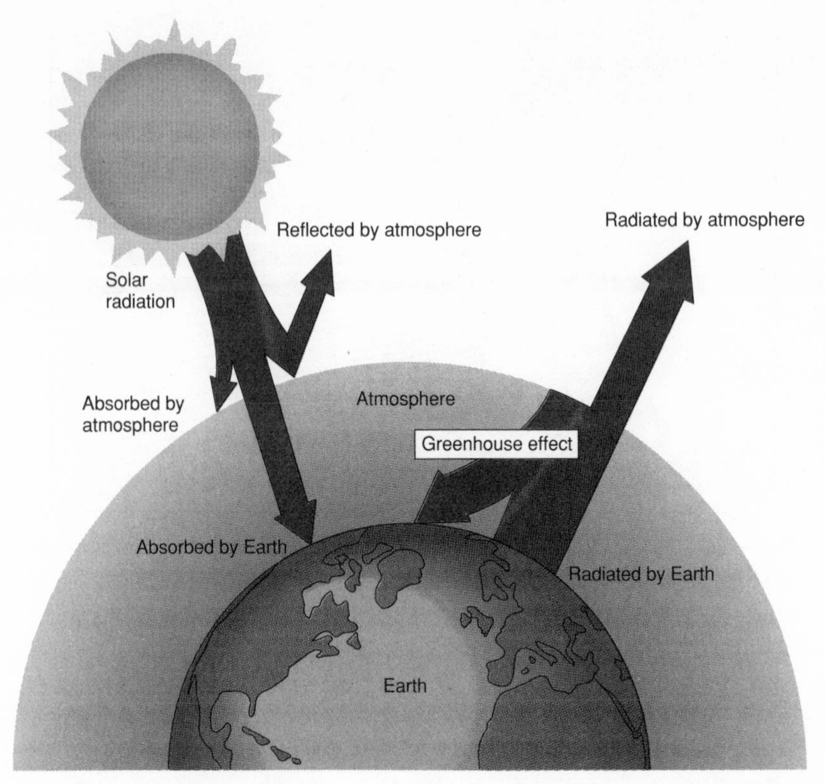

Figure 2.3. Distribution of the Sun's energy intercepted by the Earth. (Reprinted with permission from the American Chemical Society.)

contact with a cold one cools off so if no other effect intervened we would quickly be frozen because of the constant loss of energy to outer space. What saves us is the existence of greenhouse gases in the atmosphere, such as carbon dioxide, methane, and water vapor. These gases are responsible for returning more than 84 percent of the energy reflected by Earth's surface back to Earth. These gases are transparent to the sun's light, which passes through the atmosphere thus reaching Earth. Upon being reflected from Earth the nature of the light changes so that it no longer can pass through the greenhouse gases. In fact the latter now absorbs and reradiates the light energy back to Earth. The effect is similar and derives its name from the more familiar effect of light streaming into a greenhouse through transparent glass. But when reflected by the vegetation in the greenhouse it changes its character and is no longer able to escape through the glass but remains inside and warms the interior. Although the greenhouse effect played an important role in making early Earth habitable, a continued increase in the amount of greenhouse gases in the atmosphere (a result of burning fossil fuels) could have disastrous effects. This will be discussed later.

We seem to have accounted for all of the energy from the sun intercepted by the Earth: 30 percent reflected by the atmosphere, 47 percent warming us, 23 percent evaporating water. Yet we have omitted the most important source of energy that makes life on Earth possible. It is believed that .02 percent of the incoming solar radiation reaching Earth is absorbed by the leaves of plants, which then convert carbon dioxide and water (by a process known as photosynthesis) into the basic foods needed to sustain all forms of animal life. We have missed this important contribution of the sun's energy because of round-off errors in stating the percentages of the distribution of the sun's incoming radiation. For example the 47 percent warming the atmosphere should really by 46.907 percent, leaving ample room for the photosynthetic component.

Photosynthesis is nothing short of a miracle. Sunlight cannot by itself cause carbon dioxide and water to combine to form the carbohydrates we consume. Chlorophyll, a catalyst in the leaves of all plants makes this reaction possible. The elucidation of the photosynthetic effect was started when Joseph Priestley in 1771, discovered that a candle burning in a closed container goes out after a short while. He also noted that flame can be revived if a sprig of mint is added just before the flame is ready to die. This happens because oxygen is the end product of photosynthesis. In 1779, Jan Ingerhorng, a Dutch physician, demonstrated that this experiment only succeeds when light is shining on the plant, proving that the life process of the plant goes on only in the presence of light and that this process produces the oxygen. Subsequent discoveries led to identifying chlorophyll as the catalyst responsible for the photosynthesis.

The photosynthetic reaction has been studied for years and is fairly well understood but cannot be duplicated in the laboratory. The light absorbed by the leaves drives a chemical reaction that creates a series of other chemical reactions that ultimately produce adenosine triphosphate (ATP), the molecule which is the universal source of energy in all living things. The plants use ATP for their own metabolic processes and with the help of the catalyst, chlorophyll, combine the carbon dioxide and water to create the sugars and starches living things need to sustain themselves. The overall efficiency of the process, the ratio of the energy value of the food created to the energy of the incoming light is only a little over 1 percent. It is estimated that 99 percent of the incoming light energy shining on plants does not go into the production of food. This is partly because plants can only use light of a small range of colors to effect the reactions that produce starches and sugars. Sunlight consists of many different colors, and plants only use a few in photosynthesis (Fig. 2.4). The useful fraction of the light that can produce food is further reduced because plants must use a great deal of the incoming

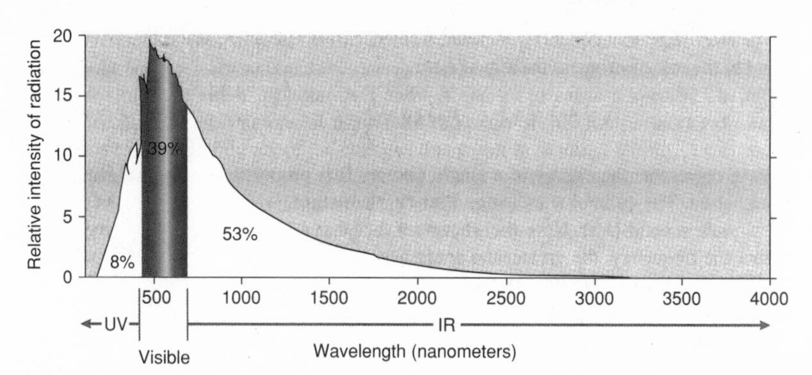

Figure 2.4. The photosynthetic process can only use visible light. This light is less than 40 percent of the light energy radiated by the Sun that reaches Earth. (Reprinted with permission from Academic Press.)

energy for their own metabolism. The prospects for duplicating photosynthesis in the laboratory with a higher efficiency seem remote at the moment.

Photosynthesis not only provides us with our food, it removes carbon dioxide from the atmosphere and supplants it with oxygen. It has been responsible for creating the raw material of what are now the fossil fuels. This miracle which originates with .02 percent of the energy of the sunlight incident on Earth and is little more than 1 percent efficient has been the mainstay of life on our planet and is at the moment responsible for about 90 percent of the energy we are using. These numbers are an indication of the enormous amount of energy the sun makes available to the Earth.

All forms of energy are convertible one to another and are subject to two physical laws which as far as anyone knows have never been violated. The first is the Law of the Conservation of Energy, which states that energy can neither be created nor destroyed when converted from one form to another. This is equivalent to a strict accounting system. Any amount of energy that comes into a system, device, or a plant must be strictly accounted for. No energy, can get lost and no additional energy can be created. The first law gives some indication that energy is a limited resource, a precious commodity for us who need it. Whatever its source, the amount that is once given is all that we can have but none of it can get lost. There is a second law further restricting how much use we can get from the energy that is available to do work. To explain this law we have to digress and say something about heat or thermal energy.

The temperature of a body is now known to be a measure of the average speed of the atoms and molecules of which the body is composed. Its thermal energy is the average energy of its constituents as measured by the average speed of a molecule times the total number of molecules in the body. Two bodies could be at the same temperature but have vastly different amounts of thermal energy. Both a thimbleful and a glassful of water could be at the same temperature because the average speed of the molecules in each is the same. However, the water in the glass will have much more thermal energy because it has many more molecules.

Until the 19th century, scientists thought that thermal energy was transported from hot bodies to cold bodies by means of a substance called caloric. The hot body would have an excess of caloric and when placed in contact with a cold body, the caloric would be transferred to the colder body, heating it up. There was something suspect about this picture due to the fact that when a body was heated its mass did not increase even though on this picture some caloric must have entered the body. It was difficult to imagine that caloric had no mass at all. This hypothesis was put to rest by Benjamin Thompson (Count Rumford), a loyalist to Great Britain during the American Revolutionary War. He noticed that cannons being bored had to be continuously cooled with cold water. He properly made a connection between mechanical energy due to work that his drills were performing and the thermal energy acquired by his cannons. The former energy was being converted to the latter. In our present picture the energy acquired by the constituent atoms of the cannons was causing them to speed up. The remnant of the caloric theory of heat is the name given a unit of energy, the calorie. This is the amount of energy required to raise the temperature of one gram of water by 1°C.

There are two sets of units which measure quantities of energy, two sets because for many years the subject was not completely understood and different views of the nature of energy led to different units for measuring it. There was thermal energy, whose unit is the calorie described above. Another unit, the joule, is used to describe a quantity of nonthermal energy, such as mechanical or electrical energy.

We inevitably have to deal with energy units. Knowing exactly the definition of these units is not as important as having some feeling for the magnitudes of the joule and the calorie. A 100-watt bulb burning for 24 hours uses almost 9 million joules of energy. One calorie is equivalent to 4.18 joules. The calorie, like the joule, is a very small unit of energy. Calories are frequently used to describe the energy content of the food we eat. Because the calorie is such a small unit, we use a unit equivalent to 1000 calories in this context. This larger unit is called a kilocalorie and written as a Calorie, with a capital C. It requires about 2000 Calories for an adult to sustain his/her life processes and to get through a typical day. This is approximately equivalent to burning a 100-watt bulb for 24 hours. Burning a ton of coal releases about 22 billion joules. When we speak of the energy consumption of countries we need a very large unit indeed. The exajoule, a billion billion joules, is such a unit. The world consumes about 400 exajoules of energy per year. To reinforce the correlation between standard of living and energy use we point out that in 1992 North Americans consumed 300 billion joules of energy per person per year whereas it was 27 billion in developing Asia and 36 billion in Africa.

In the mid-19th century, James Prescott Joule performed experiments which quantitatively confirmed Thompson's insight. As a physicist, Joule was familiar with units of energy other than thermal energy. He performed the following experiments with great precision. A paddle wheel was rotated in a container filled with water and its rise in temperature was measured. This was a quantitatively accurate duplication of Thompson's observation. He made accurate measurements of the amount of work performed. He then measured the equivalent number of calories to which this amount of work corresponded by measuring the increase of the temperature of the water (Fig. 2.5). He found despite what kind of work energy he used, electrical or mechanical, 4.18 units of work (in joules) was always equivalent to exactly one calorie. He is memorialized by having the fundamental unit of work named after him.

Before Joule did his seminal experiments, physicists had suspected that there was a Law of the Conservation of Energy but could not account for the energy that seemed to be lost whenever heat was generated in a process. After Joule demonstrated that heat was a bona fide form of energy to be considered on the same footing as all the other, such seeming violations of the first law disappeared. The energy that apparently was lost was thermal energy.

The Law of the Conservation of Energy is not the last word governing the conversion of one form of energy to another. When we slide a block across a rough table, the table heats up. The block's mechanical energy disappears into the table and warms it. No one

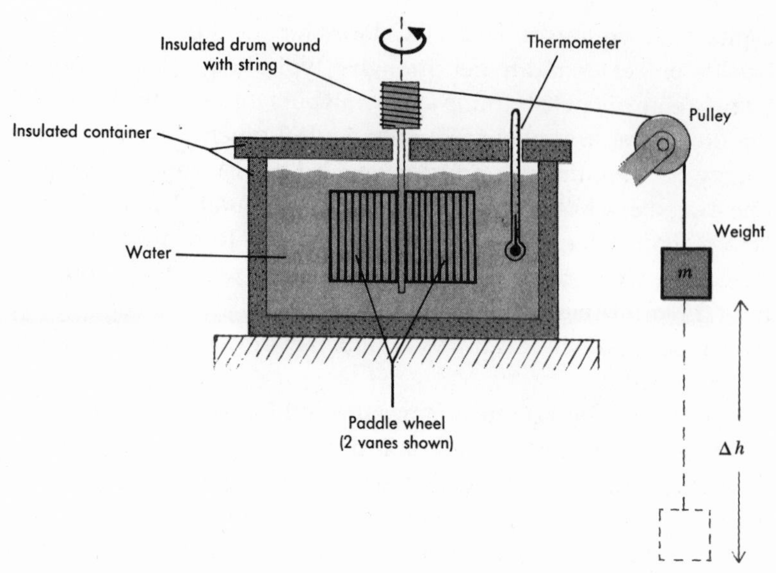

Figure 2.5. Schematic of the experiment performed by Joule to show the equivalence of thermal energy and other forms. Here, the mechanical energy of the falling weight is converted to the thermal energy of the water. (Reprinted with permission from McGraw-Hill.)

has ever constructed an engine that consists of a heavy body repeatedly rising spontaneously, cooling off and then dropping while thermal energy made available becomes completely converted into work. Such an engine would violate the Second Law of Thermodynamics. We cannot expect to see an ocean liner abstract the Ocean's thermal energy by cooling it and use the energy in an engine to propel the ship across the waters. This process would be in conformity with the Conservation of Energy, but it would violate the so-called Second Law of Thermodynamics and that is why it is never observed. This law governs thermal energy transfers between bodies and states that no engine operating in a cycle can abstract heat from a warm reservoir and convert it completely into work.

There are innumerable examples of processes similar to the one we have cited which would be in violation of the Second Law but not the first. These examples have been generalized by physicists as follows: whenever thermal energy is converted into work energy in a cyclic process (devices which do this conversion are called engines), a portion of the thermal energy becomes unavailable for doing work and remains as thermal energy in the body which does the work. Thus the conversion of thermal energy into work can never be done with 100 percent efficiency. While mechanical energy can be converted into thermal energy with 100 percent efficiency, as is the case with a block sliding on a table, no switch can be thrown in any engine to reverse the process with the same efficiency.

Despite that energy available to us is constantly being degraded into thermal energy, which cannot be completely converted into work, we seem to have plenty available to do enough work to satisfy our needs. The sun is the principal supplier of this energy. But how can the sun generate so much and how long will it last? In the 19th century physicists had access only to what was then known about possible energy sources. Their best hypothesis was that in stars in general and in the sun in particular, the gas particles that compose them attract one another to convert their gravitational energy into light, which is then radiated. Unfortunately, the lifetime of the sun as a radiating body could also be calculated according to this scenario; it turned out to be a mere 25 million years. It was known even then that the Earth was much older. So this explanation obviously failed.

Beginning with the 20th century, enormous strides were made in understanding the nature of matter. This understanding helped solve the riddle of the energy output of the sun and its long life. The first insight was that all matter consists of individual atoms and/or molecules. The Danish physicist, Niels Bohr, had provided a picture of the structure of atoms. Atoms, according to his theory, consist of electrical charges: very light negatively-charged particles (electrons) and much heavier positively-charged particles (protons). Particles with the same charge repel each other while those of opposite charge attract. There is no significance to the positive and negative designation except that just as positive and negative numbers cancel out, the effect of combining positive and negative electricity is to cancel out electrical effects. The negative charges in an atom are spread out over relatively large distances (1/100th of 1,000,000th of a meter). The protons are confined to a region called the nucleus, which is about one millionth the size of an atom.

Ordinary chemical reactions involve the rearrangement of the electrons of atoms or molecules when they interact with one another. Electrical forces drive chemical reactions. There is a different force, a nuclear force, which governs what goes on in the nucleus. It is strongly attractive at small distances. A collection of protons confined to a small region would burst apart because of the repulsive electrical force the protons would exert on one another. However, there is still another particle in all but one nucleus—the neutron. Without an electrical charge it can overcome the electrical repulsion and attract protons and other neutrons by exerting a nuclear force. When nuclei closely approach one another, nuclear reactions can occur and the nuclear particles can rearrange themselves to form different nuclear end products. In a nuclear reaction, as well as in a chemical reaction, the internal energy of the original reactants can be greater than the internal energies of the end products. The energies of the end products would then have to be greater than the energies of original reactants, according to the Law of the Conservation of Energy.

The nuclear force is 1 million times stronger than the electrical force so when nuclear reactions produce energy, the energy of the end products is likely to be 1 million times as great as the energy generated by means of chemical reactions.

We are on our way but still have not solved the riddle of the sun. Independent of the research establishing the nature of the properties of matter, Albert Einstein was formulating a theory of special relativity early in the 20th century. One of the consequences of this theory was that if a reaction occurred between masses and some mass was lost in the reaction this mass would be converted completely into energy. The conversion factor is very large—the square of the velocity of light, which is 300 million meters per second. Even a small loss of mass would result in the production of a huge amount of energy. Despite the large conversion factor when chemical reactions produce various energies (light for example), the amount of mass lost is so small that it was never connected with energy generated. However, this was not the case with nuclear reactions. In a process, the mass lost by the initial reactants was significant and could be correlated with the energy produced by a reaction.

The sun's principal constituents are hydrogen and helium. The name helium was coined when the element was first detected in the sun; the Greek word for sun is *helios*. The hydrogen and helium in the sun were identified by the characteristic radiation these elements emitted. All atoms produce a characteristic radiation when they are at a high enough temperature, and this radiation can be considered their signature.

The energy coming from the sun in the form of radiation is the result of a series of nuclear reactions. During these reactions, four protons, the nuclei of hydrogen atoms, are converted into a helium nucleus, which consists of two protons and two neutrons plus two positive electrons. The mass of the four protons is greater than the mass of the helium nucleus and the two positive electrons. The excess mass is converted into energy according to Einstein's recipe. It is a small loss of mass but the conversion of one gram of hydrogen into helium releases the same amount of energy as the combustion of twenty tons of coal. The conversion of hydrogen into helium in the sun has been an important contributor to the vast amounts of energy generated by the sun. The sun can last for a long time because it is a massive body. Every second 5 million tons of the sun's mass are converted into 30 trillion joules of energy. Despite that, astronomers have reliably estimated the sun has completed only half of its life cycle of 9 billion years. All astronomical quantities are astronomical in size.

A legitimate question in view of the Law of the Conservation of Energy is where did the energy now being delivered by the sun come from? The answer is connected to the question of how our universe originated. One theory is that 15 or 20 billion years ago, all of the constituents of the universe were created in steps after what is now called the "Big Bang"—an enormous explosion that provided all of the energy in the universe that now exists, including the energy that the sun can deliver to us.

CHAPTER 3

FOSSIL FUELS

Fossil fuels are the remains of organic matter that, over hundreds of millions of years, have undergone substantial physical changes induced by pressure and chemical changes caused by the action of bacteria. The fossil fuels are coal, oil, and natural gas. They provide the United States with about 90 percent of the energy it uses. This energy was originally provided by the sun, which made it possible for the plants to grow.

Coal was formed by vegetation that fell to the ground. We have already described how plants use the sun's energy to sustain their own metabolism and combine carbon dioxide and water to generate the sugars and starches which are the basis of our food. As plants grow, they remove carbon dioxide from the atmosphere and release oxygen. If the plants which had fallen to the ground had not been protected somehow from interacting with oxygen, they would ultimately have decayed into their original constituents: carbon dioxide and water. The presumption is that the fallen vegetation was shielded from the atmosphere by water, which was possible either because it was close to bodies of water to begin with, or because there was a sudden flooding which killed the vegetation and covered it. This permitted the process of fuel formation to begin.

One of the precursors of coal is peat, which is compacted vegetation. Peat formed at the edges of deep lakes or in swampy areas where the vegetation fell into water, thus shielding the deposits from the atmosphere. Although peat is not designated as a fossil fuel, it is the starting point for making coal and is often used as a fuel.

Coal is created from peat or other sheltered deposits of vegetable and woody matter by the action of bacteria and by increased pressure caused by upheavals of the Earth's crust. The progression of the formation of coal is from lignite, or brown coal, to bituminous to anthracite, the coal of the highest quality. Coal deposits can be as much as 400 million years old. The quality of coal is determined by its carbon content. Anthracite coal is between 90 percent and 98 percent pure carbon.

Oil has had a similar history but its ingredients are not woody plants but krill, diatoms, algae, and plankton. These, too, had to be shielded from the atmosphere probably initially by limestone and rock. A precursor for the formation of oil is organic matter trapped in shale. The organic matter in shale then turns into oil with further chemical processing made possible by increased pressure and by bacterial action. After the oil is formed, it seeps into high porosity rocks, where it is found today. Many oil deposits have been capped by domes formed by Earth tremors.

Natural gas, which consists principally of methane mixed with some ethane, propane, and butane, is found in coal mines and in oil wells. Presumably it is a byproduct of the chemical reactions that created these fossil fuels. There have been conjectures that this natural gas was part of the mix when the Earth was formed. Natural gas is responsible for many tragedies in coal mines because of its tendency to explode spontaneously. Modern practice is to pump the methane from the mines if there is enough available to market as fuel. Precautions are taken to prevent explosions when there is not enough gas in the mine to warrant trying to sell it. Still accidents occur occasionally despite these precautions.

Natural gas also occurs in oil deposits, especially those that have been capped by domes. The gushers sometimes seen when an oil well is brought in are due to natural gas in the dome escaping after the dome is pierced. But there are natural gas deposits neither in coal mines nor in oil wells whose origins are not well understood.

Even before fossil fuels became the energy darling of our civilization, the products of photosynthesis were the mainstay of living things. Primitive humans consumed the equivalent of about 2000 calories per day, about 1500 to carry on bodily functions and about 500 to do the work necessary for the activities of daily living. They directly obtained these calories by eating the edible vegetation, or indirectly by eating the animals that ate the vegetation. With the discovery of fire and the domestication of beasts, the daily use per person probably jumped to 12,000 calories per day. As the centuries passed new technologies brought more sophisticated uses for wood and then for other fuels. The Industrial Revolution upped the daily usage to an estimated 70,000 calories. Since then, further increases have been made to accommodate the sophisticated machinery that make our lives more comfortable, leading to still further increases to our caloric needs. The average daily use per person in the United States is now approaching 300,0000 calories. It is sad to note that in some countries the energy use and lifestyles remain about what they were at the beginning of the Christian era.

Population created an increase in uses found for wood. This caused the rate of wood's use to outstrip its rate of growth, with the consequent loss of forests. A large part of Europe and the United States have been deforested. Wood was the principal energy source of the world until 1890 when it was supplanted by coal. Coal maintained its preeminence until 1940 when oil took over. As we have previously noted, about 90 percent of the energy used in the United States is supplied by fossil fuels. About 40 percent comes from petroleum products and about 25 percent each from coal and natural gas.

It is easy to understand the popularity of the fossil fuels. These are energy sources which are compact, making them relatively easy to transport. Nature has already taken the giant step in creating the product. There is little manufacturing cost for oil, only the cost of pumping it from the ground, transporting it to its ultimate destination, and separating it into kerosene, gasoline, and so forth. For coal and natural gas there is only the cost of recovery and transportation. The assured market for fossil fuels has led to heavy investment not only in the infrastructures for recovery and delivery but also for the production of the devices necessary for their use, such as engines, vehicles, heating units and so forth. Consumer cost has been so low that there has been a tendency to overlook the disadvantages of these fuels, to regard them as cheap, clean, and safe—a precious "gift from God" to Earth's inhabitants.

Recently, questions have been raised about the "gift from God." The first consideration is that these are a limited resource. While it is probably true that the processes that created these fuels are still going on, the consumption far, far outpaces the production as it did with wood generations ago. After all, it took hundreds of millions of years to create the stockpile that we now enjoy. As the fuels disappear the prices inevitably will rise and the competition for them will become bitter. It seems prudent to use the window of opportunity we now have when they are plentiful to pursue strenuously the research and development of alternatives.

Fossil fuel deposits are widespread on the planet but not evenly distributed. More than half of the oil reserves are in the Middle East. Most of the natural gas reserves are in Russia. The United States, Russia, and China have over 80 percent of the coal reserves. The United States, the largest consumer of energy in the world, is particularly vulnerable to shortages in oil and must expend billions to ensure a steady adequate supply.

In the meantime, the most serious indictment of fossil fuels is the harm they do to the environment. The photosynthetic process incorporated carbon dioxide into the fuels. Along the way, sulfur, nitrogen compounds, and other hydrocarbons, some of them carcinogenic, have been incorporated as well. When the fuels are burned these substances are released into the atmosphere. The sulfur and nitrogen compounds have been costly for forests not only in the United States but in Europe. These compounds (precipitated as rain) have spoiled many of the streams and lakes in the United States and Europe. Of all the fossil fuels, coal is the most damaging, oil less so, and natural gas the least harmful to the environment.

Even the least harmful of these fuels, when burned, release into the atmosphere the carbon dioxide incorporated into the vegetation from which these fuels were formed hundreds of millions of years ago. This is a pollution which potentially can alter our lives in serious ways.

We have previously described the greenhouse effect as occurring because the atmosphere is transparent to the light coming directly from the sun but opaque to the light reflected from Earth. The opacity of the atmosphere to this reflected light serves to heat it, thus increasing Earth's average temperature. There are other gases which also contribute to the greenhouse effect but the real villain is carbon dioxide, which is always emitted when fossil fuels are burned. Of all the greenhouse gases, we have the best chance to control its concentration in the atmosphere by limiting the burning of fossil fuels.

Since about 1950, the scientific community has been concerned about the dangers of global greenhouse warming due to the increase of carbon dioxide in the atmosphere. Serious irreversible consequences can result unless we begin to replace fossil fuels with alternative renewable sources of energy very soon with the objective of eliminating them altogether.

The idea that carbon dioxide might serve as a thermostat for the atmosphere was first put forth by the Swedish chemist Svante Arrhenius in 1896. This warning stirred neither the interests nor the fears of either the scientific or lay community at that time.

There were many reasons for this disinterest. Water vapor is a greenhouse gas of much higher concentration than carbon dioxide in the atmosphere. It was thought that burning fossil fuels added so little additional greenhouse gas (carbon dioxide) to the atmosphere to have an effect. At that time measurements of the amount of carbon dioxide in the air were not very accurate nor were they taken frequently enough to measure whether it was increasing. Carbon dioxide was considered a slight impurity in the atmosphere, about 30 times less abundant than argon. The ocean was known to contain 50 times as much carbon dioxide as the atmosphere so it was thought it probably could serve as a reliable regulator of the atmospheric concentration. Arrhenius was probably ignored because, during the late nineteenth century, it was inconceivable even to the scientific community that human activity could transform the face of Earth for the worse.

After the end of World War II, all of the evidence that might convince anyone that the effects of increased carbon dioxide in the atmosphere were benign was overturned. It turned out that water vapor and carbon dioxide did not block exactly the same radiation, so carbon dioxide could have an important effect; the ocean did absorb some carbon dioxide but did not retain much of it; carbon dioxide levels in the atmosphere were measured accurately and correlated with an increase in average temperatures in the world. By 1988, James Hansen, director of the National Aeronautical Space Administration's (NASA) Goddard Institute for Space Studies, testified before the United States Congress that greenhouse warming could be upon us. He pointed out that the unprecedented increase in carbon dioxide concentration in the atmosphere from 280 parts per million (prior to the Industrial Revolution) to about 350 parts per million (at the time that he testified) was cause for concern. Also the 1980s was the warmest decade ever recorded; this might be signaling the onset of the warming of the Earth with potentially serious environmental consequences.

There have always been large fluctuations in the Earth's temperature year by year (Fig. 3.1). The heat waves of 1980s were not a sure sign that the Earth was seriously warming up. However, by 1988, technological advances in computing had provided facilities for creating projections of how rapidly the climate might change and what the consequences of climate change could be. These projections were quite thorough and took into account positive and negative mechanisms for temperature changes: the effect of greenhouse gases but also the more rapid rate of photosynthesis which would remove the carbon dioxide; the reduced reflections of the light from Earth because there would be less snow and ice as the temperature increased; and the effects of increased cloud cover. The conclusions of the projections were that if we maintain our present rate of emission, the carbon dioxide concentration will double by the middle of the next century and the average temperature might increase by as much as 4.5–6.0°F.

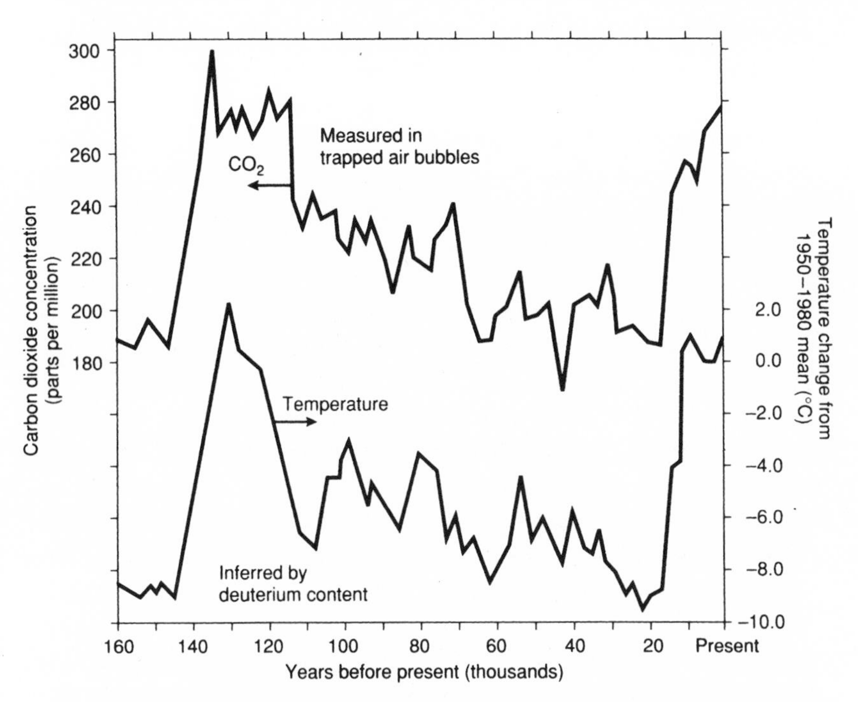

Figure 3.1. Despite substantial, irregular year to year fluctuations the temperature variations of Earth's atmosphere correlate well with the carbon dioxide concentrations for the past 160,000 years. The data after 1980 would show an increase in both. (Reprinted with permission from the American Chemical Society.)

While this might not seem like a very large increase in the average temperature the consequences could be very severe. Even during the last Ice Age, Earth's average temperature decreased by only 5°F. It could cause sweeping changes in patterns of rainfall and drought. It would be highly disruptive to ongoing agricultural practices. It would affect in an unpredictable way, the plant systems of the Earth, including forests. It would play havoc with coast lines, especially dangerous for small island nations, because of the flooding accompanying the melting of polar ice, something that has already begun to occur. Economists estimate the annual losses might be untold billions of dollars.

When this problem was brought to the attention of the United States Senate by a respected senior scientist it immediately got the government involved in sponsoring climatic research. In 1994 $1.4 million was appropriated for this work. The United Nations organized an Intergovernmental Panel on Climate Change in 1988. These efforts also led to the organization of an International Conference on Climate Change held in Rio de Janeiro in 1992. This meeting resulted in a treaty signed by 106 nations designed to return the emissions of greenhouse gases, especially carbon dioxide, to the values of 1990 by the year 2000. In 1995 the Intergovernmental Panel on Climate Change, which had been organized by the United Nations, issued a report stating that greenhouse warming is a real effect and may already be upon us. Still the pledges made in Rio de Janeiro have been honored universally in the breach.

At present, almost all scientists agree that global greenhouse warming due to increased concentrations of carbon dioxide in the atmosphere is a bona fide effect and greenhouse warming has begun. While there is no agreement that the effect has been established beyond any doubt there are indications that something serious is going on. Dr. Ranga B. Mynen of Boston University reports in the scientific publication, *Nature*, that between 1981 and 1991 there was an increase in the peak growing season north of the 45th parallel. Spring is arriving earlier and Fall is departing later. The wild flowers in the Alps are now growing at a higher elevation than they had been. Island nations have reported that their coast line is diminishing due to flooding, which does not recede. Mr. Ross Gelspan, in his book *The Heat is On* (Addison Wesley Longman) reports that the insurance industry paid out less than $2 billion a year for weather-related damages in the 1980s but between 1990 and 1995 they paid out $30 billion a year. There are other indications, none leading to an unequivocal conclusion that weather-related data are changing.

It should come as no surprise that there would be a psychological tendency to do nothing about global warming despite these indications. The problem is complicated and the consequences cannot be predicted with absolute certainty. But faced with a potentially drastic outcome that is irreversible when it is really upon us, prudent policy would dictate that we prepare for the worst. This is especially true since the dynamics of meteorology are not very well understood but there are indications that small temperature changes in the surface of the oceans could have serious destructive effects. There are uncertainties. Even if greenhouse warming should turn out to be a false alarm, the introduction of alternative energy sources will not have done any harm. In fact, would remove the specter of shortages of fossil fuels. If it is a real danger, doing nothing until all uncertainties are removed could cause a great deal of pain.

While the vast majority of the scientific experts agree with the United Nations' Panels conclusions, a handful of academics, some in respectable schools, have publicly said such an effect does not exist and have attacked these conclusions. Mr Gelbspan, in his book, points out that despite the small number and minority view of the academics, they have had a disproportionately large influence. They are given forums by the Legislature to testify; their opinions influence Congress. For example, Dr. S. Fred Singer, a gadfly scientist, not listed in the American Men and Women of Science and who does not publish in refereed journals has been repeatedly given an opportunity to testify that global warming is a "media scare" by Congressman Tom DeLay (R-TX). He also points out that they have been able to attract the attention of the media way out of proportion to their number or standing in the scientific community.

This has the potential to be an international problem with serious political and economic consequences. At present, 75 percent of the planet's population consumes only 25 percent of its energy. There is undoubtedly an expectation that the poorer 75 percent will want to increase its percentage of energy usage using inexpensive fossil fuels with consequences for an intensified greenhouse effect. It is also true that in developed countries there is an enormous financial investment in the use of fossil fuels. Resistance can be expected to occur, in fact has already occurred, to changing over to energy sources using different technologies.

In late 1997, 160 nations met in Kyoto, Japan to attempt to address the problem of greenhouse warming. The good news is that this diverse group managed to hammer out the terms of a treaty, not a pledge, yet to be ratified by the participants in 1998. The bad news is that the terms agreed upon will have little effect on the potential danger. Only 39 nations of the 160 countries would have limits placed on their emissions—121 are exempt. Among the exempted countries are some, like China, that are already emitting substantial amounts of greenhouse gases, and others whose emissions are likely to grow in the future. Nations with different levels of industrial production have different limits placed on their emissions. The highly industrial ones pledged to reduce their emissions from 6 to 8 percent of their 1990 levels. Some, like Russia, are asked to stabilize their emissions to the 1990 levels by the year 2012. Still others are given targets allowing for the increase of the 1990 level of emissions.

The agreements reached in Kyoto are but a small step in addressing the potential problem. Even if the plan were implemented immediately a great deal of harm will already have been done because of the emissions that have occurred to date. It is not certain that even this small step is assured. The United States Senate is already on record as saying they felt we should adopt a treaty that does not require "meaningful participation" by developing countries. Apart from this opposition the propaganda machines of those whose ox may be gored seems to begin full swing. On the one hand, one hears the old refrain that greenhouse warming is not a real threat but only an artifact of a computer program, on the other, the dire economic hardships that would be delivered to us should anything be done to curtail the use of fossil fuels.

Even the modest agreement proposed in Kyoto is being attacked. Many members of the scientific community have been sent a packet that includes an imitation of a reprint that looks like an article that might have been published in the *Journal of the National Academy of Sciences*. But this article has never been published in any refereed scientific journal. The authors list themselves as belonging to two organizations, "The Oregon Institute of Science and Medicine" and the "George C. Marshall Institute," neither of which is well known scientifically. The point of the article is that increased carbon dioxide in the environment is good for us. It makes all vegetation, including flowers, grow faster. The authors neglect to point out any of the hazards which the scientific community has been warning against including the fact that a good part of the turf might be under water. The mailing was accompanied by a request to sign a petition urging that no action be taken to limit the use of fossil fuels. Sadly, the petition and the article was endorsed by Dr. Frederick Seitz, a distinguished physicist, former president of the National Academy of Sciences, and president emeritus of Rockefeller University. Dr. Seitz's previous campaign endorsing public policy was to urge the funding of the Strategic Defense Initiative (Star Wars).

There are ethical reasons for assuming the worst case scenario and taking some action. Should this generation burden future generations with a problem which, if it exists, might have no resolution once it has already done its damage? It is a question of what responsibility individual members of a society may have to its well being.

Adam Smith in his book, *The Wealth of Nations*, popularized the idea that an individual who "intends only to his own gain" is "led by an invisible hand to promote ... the public interest." An article in *Science* entitled "The Tragedy of the Commons" by Professor Garrett Hardin (University of California Santa Barbara) takes the opposite view. The word "tragedy" in the title has the meaning originally proposed by Alfred Norton Whitehead: "the essence of tragedy is not unhappiness. It resides in the solemnity of the remorseless working of things."

To illustrate the tragedy, Hardin uses the model of a pasture where the grass is available to all. If the use of the pasture by humans and their cattle are well below the carrying capacity of the land no tragedy ensues. Once this limit is exceeded the inherent logic of the situation generates the tragedy. Each herdsman, if he is rational, will add as many beasts to his flock as he possibly can. Each animal benefits him alone because he alone can profit from its sale. However, the down side, the peril of overgrazing, is shared by all members of the community. The conclusion reached by any individual herdsman is the conclusion reached by everyone. Thus each person pursuing his/her own best self interest serves to ruin the community as the commons becomes overgrazed and useless.

Lest one think that this is an artificial example, in 1994 we had a real life drama which precisely illustrated this principle. The Georges Bank fishing grounds, the continental shelf east of Cape Cod, can be thought of as the common grazing ground. The New England fishermen over the years added boat after boat to their fleets to increase the catch and thus increase their income. In 1994 the government had to restrict the fishing there to reduce the catch by over 50 percent. The stocks of cod, flounder, and haddock had become so scarce that a danger loomed that they would not be able to reproduce and these once fabulously rich fishing grounds could be lost forever.

The moral of the "Tragedy of the Commons" is that when it comes to air and water, the parts of our environment that cannot be fenced in, each person has the obligation not to do what is best for him/her, but has a responsibility to do what is best for the community. If such rational behavior is unlikely, it is the responsibility of the government to create incentives to direct people toward socially acceptable behavior.

Unfortunately, at the moment it seems as if the mood of the United States population is sympathetic to the free market economy as expounded by Adam Smith, rather than to trying to save the "Commons." As noted earlier, a measure was introduced in the 1994 Congress to impose a substantial tax on the use of gasoline, a fossil fuel. Instead, a tax of about 4 cents a gallon was enacted, much less than the fluctuations in price during the course of the year and not enough to cut down the use of gasoline. During the presidential campaign of 1996, one of the candidates urged the repeal of even this modest increase.

Professor E. O. Wilson, of Harvard University, the world's preeminent authority on ant populations, has pointed out that the ant species has existed for hundreds of millions of years. He attributes this longevity to the fact that ants are programmed genetically to behave in ways to benefit the ant society rather than the individual ant. Could Homo Sapiens, which is only a few hundreds of thousands of years old, use its genetically programmed intelligence to make its species behave more like the ant's, to give it a chance for similar longevity?

CHAPTER 4

OIL

The motor car is one of the greatest inventions of all time in its effect on society. Although there have been attempts to power automobiles using coal as a fuel (the Stanley Steamer was the most famous car of that genre), soon engines that used fuel derived from petroleum drove the coal powered ones out of the market. They were lighter, more efficient, and generally superior to steam engines. This marriage of motor cars and petroleum products led to oil becoming the most used fossil fuel. Oil's use in generating electricity and its effectiveness in powering airplanes, locomotives, and speedboats have also helped to put it ahead of all other sources of energy. It has occupied this premium position since 1940 and will probably continue to do so as long as ample supplies of oil are available. As had been noted, it supplies 40 percent of the energy used in the United States with coal and natural gas each contributing 25 percent.

Oil has been known since very ancient times. People first became aware of it when it appeared in certain rock outcroppings. Its name petroleum (petro- rock-, oleum- oil) is derived from rocks impregnated with oil that could support a flame. These were frequently found near places where oil had oozed out of the ground.

The following information about the early history of oil is taken from a book written by Abraham Gesner in the middle of the nineteenth century and reprinted in 1968 by A. M. Kelley, entitled *A Practical Treatise on Coal, Petroleum and Other Distilled Oils*. Herodotus, the Greek historian, mentions that Anderrica wells yielded oil among other products. Oil that had oozed out of the ground could be purified by pouring it into a container and leaving it alone for a while. Pure oil would rise to the top of the container and could be drawn off. The primary use of this product was for illumination.

In 1694, Eeele, Hancock, and Portlock obtained a patent for extracting pitch, tar, and oil out of an oily rock. Beginning in the 18th century, oil was obtained from coal by distilling it in the absence of air. In 1846, Abraham Gesner, the author of the treatise, was in the business of extracting oil by distilling coal. This oil was used for lighting, caulking, waterproofing, lubricating (especially in axle grease) and dressing leather. It was also used in varnishes and paints and for medicinal purposes. After the discovery of oil wells, the industry based on using coal as a raw material for obtaining oil disappeared. The following is an 1861 calculation quoted by Gesner, "If the price of crude petroleum reaches a price of 35 cents per gallon, it would again be economical to obtain oil by distilling coal." We may see in this a hint of things to come when the sources of oil dry up. This fragment of early history is an indication that oil appeared early in the affairs of humans.

The origins of oil are not completely understood, but there is a consensus that the following conditions must be satisfied if an oil deposit is to be formed. There must have been a supply of organic material from either plants or animals. Oxygen must have been excluded from the environment. There must have been a supply of sediment sufficient to contain the organic matter to prevent its dispersal. There must have been bacteria present that could use the organic matter as food thus transforming it chemically. Petroleum deposits have a great deal of oxygen removed from their compounds, presumably taken by these bacteria to sustain themselves. The low concentrations of oxygen in fossil fuels are responsible for their efficiency as an energy source. It is still not clear what exactly the mechanism has been for the formation of large underground deposits of oil. A better understanding of the source would help geologists determine where to dig to find it.

The first well in the United States was dug in Pennsylvania in 1859. By 1864 there was a sufficient worldwide market for oil so that

20 million tons were shipped from the United States. The largest exploration conducted in the 19th century was in the Baku fields in Azerbaijan of the former Soviet Union. The oil from these still producing fields led to the founding of the giant petroleum companies, Shell Oil, Royal Dutch, and British Petroleum. Before the motor car age the principal use of oil was for illumination, especially in lamps. The lightest fraction, what is now its most important component, gasoline, was discarded.

The first oil wells were dug where there was surface evidence of oil in the form of seepage through the porous rock. Oil generally appears near porous rock and indeed in oil wells most of the oil has seeped into the rocks. As oil became more valuable, geologists became interested in trying to determine where it would be profitable to drill for it. In the hundreds of millions of years during which pools of petroleum were being formed, the Earth's crust underwent many distortions. One of the most important, from a oil prospecting point of view, is the anticline formation. The dome that formed during some violent compression of the crust created a shield for the well and kept the oxygen away (Fig. 4.1). There are other ways in which the wells were capped but which did not betray themselves by distortions of the surface of the Earth. For example, they might have been capped by salt deposits, or plugged by a volcanic intrusion or by other methods that geologists have learned to recognize. Eventually the geologists developed sophisticated methods for discovering where oil might be hiding. One of the most useful is seismic exploration. This involves setting off an explosion deep in the Earth and detecting the sound waves as they are transmitted through the Earth. The knowledge of the different velocities of these waves in different media enables geologists to determine likely sites for wells. It is probable that the vast majority of possible sites have already been found.

The recovery of oil from wells is done in stages. The primary recovery is to take advantage of the natural pressure of the oil, or the surrounding gas and water, to pump the oil out. When that has been exhausted, secondary and enhanced methods of recovery are used. The oil resides not only in a pool, but in the surrounding

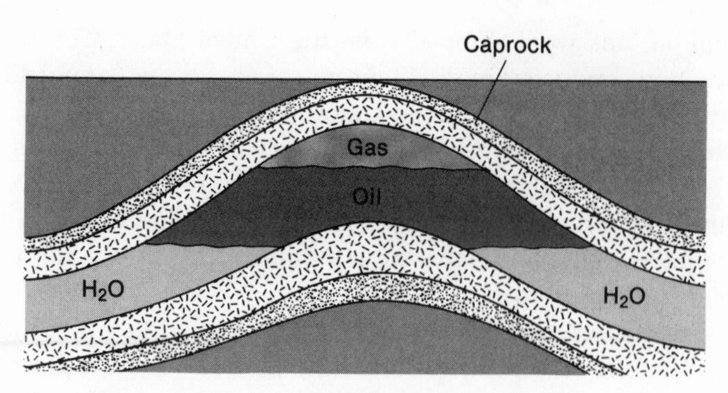

Figure 4.1. A dome created to prevent the oxygen of the air to decompose the oil. (Reprinted with permission from Saunders College Publishing.)

porous rock into which it has migrated. This oil is recovered in a variety of ways: by reducing its viscosity, especially for heavier oils, by thermal or chemical means; and by forcing the oil out of its home by injecting water, steam, gases, or chemicals. Virtually all of the wells in the United States use these secondary and enhanced recovery methods. These now account for about 10 percent of the oil produced here. Should the price of oil go up, more sophisticated and more expensive methods of recovery would undoubtedly be introduced until the cost of recovery would exceed the market value of the oil. These are unlikely to affect significantly the length of time oil will be available as an important energy resource.

Oil is such a critical source of energy for us that the question of how long it might last at a reasonable cost is an important one. Geologists attempt to do this by estimating the known reserves, reserves that have been established from wells already discovered. To be safe in estimating the oil resources, this amount is usually doubled to allow for new sources that might be found.

Some estimates made by geologists have been startlingly correct. The most ingenious was one made by the late M. King Hubbert, an employee of the United States Geological Survey. His estimate does not rely on the data of the number of reserves and the likelihood of other discoveries. Rather it concentrates on the history of the production of oil in time and takes into account the increase in price of a commodity as it becomes more scarce.

The basic hypothesis of Hubbert's estimate is that when an important commodity is introduced into the marketplace, consumption is very small and production is correspondingly small because existing technologies have not yet been readily adapted to its use. Only gradually do the consumption and production build up. Much later, when the supply nears exhaustion, a combination of increased cost due to unavailability and competition from substitutes leads to limited production because the price is now high and the market has shrunk. The curve of production of the resource plotted against time starts at a low value and ends at a low value. Somewhere in between, the curve must reach its maximum value.

The mathematicians have long been familiar with a curve that has these characteristics. It is called a Gaussian, or more popularly a bell-shaped curve (Fig. 4.2). This curve rather accurately describes a variety of things we can observe in everyday life—for example the distribution of heights among adult males in the United States, the

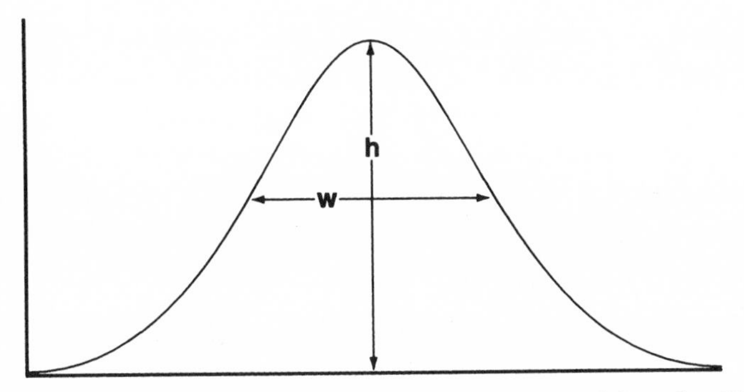

Figure 4.2. All Gaussian curves can be expressed by a general formula which has two arbitrary constants, the maximum height "h" and the width "w" of the "bell." If the value of a particular Gaussian is known for two different values, the value of the constants for this Gaussian can be determined and the entire curve is completely known.

distribution of grades on the Scholastic Aptitude Test, and many similar statistical properties. Professors are so enamored of this curve that many assign grades in some large classes based on their assumption that the achievements of the students in their classes can be represented by a Gaussian. The curve is a universal one whose formula contains only two unknown parameters. If the value of the presumed Gaussian is known for two values of the time the entire curve can be reconstructed.

Hubbert assumed that the Gaussian accurately describes the production of oil in the United States. Since he had the data for the first part of the curve, he could reconstruct the rest because that fixes the two unspecified parameters that determine the curve. Fig. 4.3 illustrates the curve predicted by Hubbert and the actual production. It shows he predicted the maximum production in the United States would occur in 1971, a prediction made long before it actually happened in 1970. The curve also displays the known reserves available

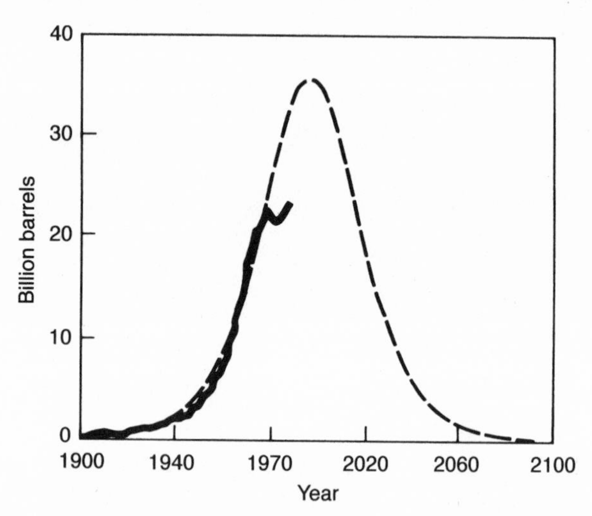

Figure 4.3. The actual production of oil (the dotted curve) in the United States closely conforms to Hubbert's hypothesis (the solid curve). (Reprinted with permission from Saunders College Publishing.)

and estimates the amount of undiscovered resources still to be found. Since the 1970s the United States has been importing more oil than it has been extracting from native sources.

Many other estimates using different methods have been made to assess how much oil remains to be exploited and used. One method compares the current number of unsuccessful attempts to locate a well before it is found with how many attempts it took in years past. There are variations of Hubbert's analysis made for explorations in other parts of the world. In addition, of course, estimates of the known reserves and the likelihood of increased discoveries are also used. All of these estimates, while they do not agree in detail, all come to the same conclusion: petroleum is definitely a limited resource. It is difficult to say how long petroleum will last because that estimate depends on its price, the success of conservation methods, changes in technologies both for prospecting and for use. There is no doubt that at the present rate of use the United States supply is likely to last about 30–50 years and the worldwide supply in the neighborhood of 100 years. There have been some spectacular new sources discovered recently but the amount of oil discovered in any year is smaller than the increased amount used in that year over the previous year.

It is easy to see why oil has become the preeminent fossil fuel. Of these fuels oil yields the most energy per unit volume. Also, very important, it is easily transportable. Pipelines have been built all over the world to move oil from the fields where it is extracted to ports whence it can be shipped everywhere. A profitable industry has grown up to do this work. Moreover the market price of oil is very low for the amount of energy it can deliver, low because the price includes prospecting, extracting, and shipping but excludes extensive manufacture. The refining of petroleum not only yields many useful fuels but also many chemicals which are root stock for modern plastics and medicines. It seems really to be "God's gift" to humankind. Unfortunately, if we build a worldwide economy on oil's availability and it disappears, that wonderful gift may become a curse.

If one were to use oil solely as a root stock for complex chemicals that could be fashioned into socially useful products, its lifetime could be increased and its environmental impact made less burdensome. Dimitri Mendeleev, the Russian chemist who discovered the periodic table of the elements, remarked when inspecting the oil fields of Pennsylvania that it was criminal to waste so valuable resource as oil by burning it.

On the way to creating the oil more or less easily extracted from wells nature created some precursor oil-bearing rocks, called shale. There are enormous deposits of oil shale in the United States, largely in the states of Colorado, Utah, and Wyoming. The amount of oil in the shale is estimated to exceed all of the known oil reserves in the entire world. A problem, however, exists in recovering this bonanza and became apparent in the 1970s.

When OPEC unexpectedly caused the price of oil to jump from under $10 a barrel to over $30 in 1973, the United States government thought it prudent to invest in methods to recover oil from shale. The economics of such a project were so costly, and its chances of success so remote, that no private company could afford to undertake it. It was clearly a case where subsidies from the government were justified by the potential usefulness to society of a successful outcome.

The oil in shale is contained in a substance called kerogen which is tightly bound to the rock. This rock can be excavated and then processed above ground. Or, if shale is close to the surface, it can be strip mined by first removing the topsoil to expose the rock and then using heavy machinery to excavate the shale.

The oil is extracted from the rock by heating it to about 400°C and then drawing off the oil. This leads to one serious environmental problem. The density of the spent shale is about ⅓ less than the original rock. Heating the rock leads to a "popcorn effect." If the shale had originally been strip mined the spent volume of residue is too great to be returned to the hole whence it came. If the shale had been mined from the deep, the problem of getting rid of the spent shale is even harder to solve. One method of disposing of the residue is to process it into a fine powder to be disposed of above ground. It is estimated that for one million barrels of oil, 1.3 million tons of shale would have to be disposed of. This might be dropped into canyons. If this were done, processing enough shale to produce 1 million barrels of oil would create a foot-high residue in a 100 by 300 meter cavity—after washing the shale down with 3 million barrels of water. The daily consumption of oil in the United States is about 15 million barrels. That pile of spent shale would rise 15 feet a day in such a cavity and 45 million barrels of water would be needed to wash it down.

Shale is found in some of the most scenic states, which also have the problem of potentially severe water shortages. Even if the water could become available it is doubtful that the residents of these states would welcome an activity that would place such an environmental burden on them. In addition, the areas to be exploited are also attractive to tourists. The building of mining towns where natural beauty now prevails is not a prospect that neither the tourists nor the residents of these Western states would relish.

The recovery of oil from shale seemed promising because a similar industry, the recovery of oil from tar sands, was a commercially successful business in Alberta, Canada. But the process of recovering oil from tar sands was actually much simpler than getting the oil out of kerogen. The oil in tar sands is in the form of bitumen. The sands can be floated on water, the bitumen separated out, and the oil recovered from the bitumen by chemical means rather than by heating—a much simpler process. Canada is also fortunate the tar sands are not found in scenic areas. In any case there are not enough tar sands in the United States to make any dent on satisfying oil consumption here and not enough tar sands in the world to compensate for the disappearance of the present sources of petroleum.

We have concentrated on the potential scarcity of this important energy source. There are some serious down sides to its use already alluded to: the sulfur and nitrogen compounds and carcinogenic hydrocarbons released into the air when it is burned. Above all, as we have indicated, no matter how many of the pollutants are removed, there is still the carbon dioxide emissions contributing to the greenhouse effect.

The most important energy source underpinning our way of life will be in short supply within two or three generations. Shortly we shall see there are promising alternatives, but these require some nurturing. The replacement of oil by such alternatives will require time and a commitment from the entire community. The business interests in oil are well entrenched and politically powerful. The alternatives require, for the most part, industries to be established that will probably require some initial governmental subsidies to enable them to compete. One can hope this transition will be effected without wrenching political battles. Since the use of petroleum is worldwide, we should hope too that the transition could be effected without any military battles as well. The key to a successful peaceful transition is the participation of a concerned public in a serious effort to make the need for substantial early investment in alternative sources of energy recognized.

CHAPTER 5

COAL

Of all the fossil fuels, coal is the most plentiful. It is also the most damaging to the environment. When beds of coal were being formed, Earth must have been a vastly different place from what it is now. Coal is found in the most unlikely places, including Antarctica. Probably it was not formed there, so there must have been some serious upheavals on Earth to transport the deposits to that region, so inhospitable now to growing vegetation.

Coal is widely, but unevenly, distributed globally. The United States, Russia, and China have about 80 percent of the known deposits. The amount of coal available is known better than the stores of the other fossil fuels, oil, and natural gas. It is presumed most of the deposits are now known. Their capacities have been measured by digging holes into a deposit some distances apart and estimating the quantity of coal in that deposit. In estimating coal reserves, the experts double the amount of known reserves and then add some factor for future discoveries. The amount so estimated is not a measure of how much can be mined economically, however. For economic recovery the seam must be at least 60 centimeters (24 inches) thick and no deeper that 2000 meters (6500 ft.), unless the coal can be recovered by strip mining, to be discussed later.

Despite the uncertainties in these estimates one can say that enough coal is available to last for several hundred years at the present rate of use. It could probably last for over 100 years if it were the sole source of energy at the present level of use. Dependable estimates are difficult to make, however, because many things could change. Among these are the level of use, conservation, price changes, technological improvement in mining, changes in the law. All these and other considerations must be noted before a true estimate can be given. Nevertheless we can say that although coal is definitely a finite energy resource there is plenty available at the moment, but probably not so much that we can become complacent and not seek other renewable sources.

As early as the 15th century, coal was used in England to evaporate sea water by heating it for the salt. Soon coal was used in a more sophisticated way to make coke, which was then used to smelt iron and convert it into steel. To make coke, coal is heated to a high temperature in the absence of air so that all of the volatile components are driven off. What is left is practically pure carbon without the pollutants, sulfur, nitrogen compounds, and hydrocarbons. Coke is combined with iron to make steel.

The volatile residue when coal is converted into coke contains extremely valuable chemicals. The nitrogen compounds especially can be used to make ammonia, the versatile starting point for many useful modern materials. Mendeleev's reaction to the burning of coal was similar to his reaction to the burning of oil, that it was criminal in view of the valuable chemicals it contains that can serve as a root stock for many useful compounds to burn it.

Today, in the age of plastics, synthetic fibers, and exotic dyes all based on the easily derived and thus cheap hydrocarbons in coal, it is even sounder advice. However, converting all coal to coke and then using coke as a fuel would increase its price enough to make it unattractive in the market place, even considering the income from the sale of the residues.

Coal is a very versatile substance. The least expensive way to use it is to burn it as it comes from the mine. But for some particular uses, it can be converted into many different products. Prior to World War II coal was extensively used to create a gas that could be used for cooking and heating and for illumination. By heating coal to a moderate temperature in the presence of steam, it can be converted to a combustible mixture of carbon monoxide, hydrogen and methane.

This so-called "water" gas was once an important commercial product. It was replaced by natural gas, which delivers about twice the energy per cubic foot at a lower cost. The "water" gas used to be stored in "gas-works" in the seedy part of cities and towns. (At one time the St. Louis baseball team was affectionately called the Gas House Gang to emphasize its humble origins.) Natural gas may give out before coal so one could look to coal to supply a substitute for its use. Apparently, with this in mind, the United States government maintains an experimental station in North Dakota to find inexpensive substitutes for natural gas based on coal.

Coal can also be converted into a liquid fuel to serve as a substitute for gasoline. This is done by a chemical process known as the Fischer–Tropsch synthesis and involves heating it with hydrogen, carbon monoxide, and a suitable catalyst to a high temperature.

While this process produces liquid petroleum fuels suitable to drive engines the product is much more expensive than gasoline and is very polluting to the environment. Still there have been commercial installations manufacturing this substitute. South Africa, which has no indigenous petroleum source, is short of foreign capital, and cannot afford to purchase much oil from abroad, uses this method to produce the vehicular fuel needed for its automobiles and trucks.

During World War II, Germany, which did not have access to enough oil, used this process extensively to fuel its war machine. In 1944, Germany was able to manufacture 5.5 million tons of synthetic gasoline using coal as a base.

One principal disadvantages of coal compared to other fossil fuels is that it is not easily transportable. Worse, it is mined in places far from where it is used. Attempts to solve this problem have had a mixed success. One scheme was to create a slurry of crushed coal and water to be delivered by pipes to where the coal is to be used. Our serious potential shortage of water in the United States and in other parts of the world means there is probably not enough water available for slurried coal to make a serious contribution to our energy needs.

Another avenue of research has been to convert coal to gas underground before it is mined and then distribute the fuel as obtained.

Gasification schemes have been technically successful but less so commercially. This technology does not have the benefit of a network of pipelines to transport the gas.

The Clean Air Act stimulated research efforts in the United States to make coal's usage more efficient through gasification or other means and to make its environment in fact more benign. There may have been a decision at the highest levels of government because coal is the most plentiful fossil fuel and likely to be around for the longest time, that energy policy should attempt to make its environmental impact as benign and its use as efficient as possible so it will last.

Part of the research activity has been to eliminate the harmful emissions produced when coal is burned. The most serious pollutant is sulfur. Scrubbers have been invented and are used to remove part of the sulfur from the coal. A scrubber is a limestone-lined smokestack. Limestone chemically combines with sulfur, thus extracting it from the smoke. But this only removes part of the sulfur. A supplement to scrubbers is to remove the sulfur impurities prior to scrubbing. The sulfur in the coal is in the form of pyrites, which can be removed mechanically from coal once it has been pulverized. The combination of these two methods removes practically all of the sulfur. Both methods are still being improved.

The most serious research direction involves creating more efficient end uses for coal. Most coal is used in electrical generating plants. The coal is used to heat a gas to drive an engine that is used to drive a turbine that drives an electrical generator. A turbine is a device which converts rectilinear motion into circular motion. In the last century Michael Faraday in England and Joseph Henry in the United States showed that an electric current can be created in a wire by moving it across a region where magnets have created a magnetic field. The electrical generator is a complex engineering device whose simple realization is shown in Fig. 5.1. In a practical generator an engine drives a turbine that causes a magnet to rotate. Wires are wound around the magnet in an external magnetic environment. This creates a current in the wires which is then sent to all users.

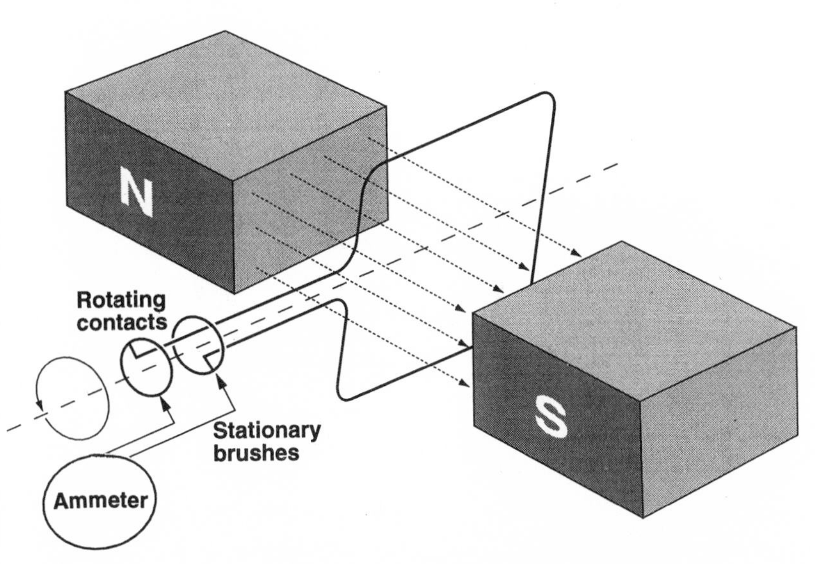

Figure 5.1. Principle of conversion of mechanical energy into electrical energy discovered by Michael Faraday and Joseph Henry. A loop of wire rotating near a magnet will produce a current in a wire connected to the loop.

The electrical generator itself is already very efficient so the research efforts under the Clean Air Act have concentrated on improving the efficiency of the engines which drive the turbines. Engine efficiencies, that is the ratio of energy put in to the energy produced, are notoriously small. Improving engine efficiency is important not only in the context of electrical generation but in many other problems associated with energy use, such as engines used in transportation. Billions of dollars have been allocated for this effort both by the government and by industry.

It may be of some interest in view of the importance of the problem of engine efficiency to understand why it is so low. In Fig. 5.2 we represent schematically the operation of an engine in which an amount energy in the form of heat Q1 is converted to an amount of

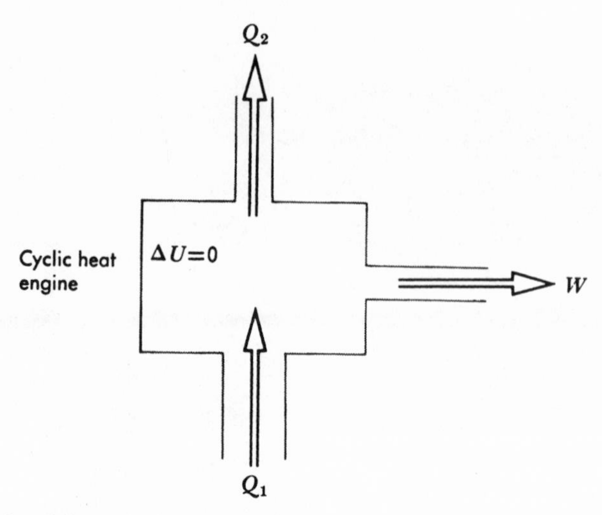

Figure 5.2. Schematic demonstrating the energy flow in an engine during one cycle. (Reprinted with permission from McGraw-Hill.)

work W. In an automobile engine, the input heat is that generated by the burning of the mixture of gasoline vapor and air after it has been ignited by a spark. At the end of the cycle, whatever remains in the piston is expelled into the air. The expelled gas has a finite amount of energy $Q2$ because of the energy of its individual molecules. The input energy is therefore not completely converted into work because of the energy $Q2$ which is not converted to work. Thus the efficiency is adversely affected.

If $Q2$ were equal to zero we would have a 100 percent efficient engine in principle. The laws of physics indicate this could only occur if the gas were expelled at the temperature of absolute zero: −273.15°C. The laws of physics also inform us that nature has conspired not to permit such a temperature to be reached.

In 1824, the French engineer Sadi Carnot showed the maximum possible efficiency of engines, neglecting the mechanical and frictional losses, depends on the ratio of the difference between the highest absolute temperature attained in the engine to the absolute temperature of the expelled gas (absolute temperature of a gas is its Celsius temperature plus 273.15 degrees). Modern research in engine design attempts to maximize the temperature difference referred to above and to minimize the temperature of the expelled gas. Engines with efficiencies of 50 percent or higher are considered state of the art.

We seem to be committed to a future with coal. Its versatility and its abundance means it could very well satisfy our energy needs for some time. Despite efforts to clean up its pollution there is still one important environmental problem attendant on its use that cannot be overcome. However we clean coal, when it is finally burned it releases carbon dioxide, the greenhouse gas likely to cause the most serious problems. And we have already described the consequences of not reducing the amount of carbon dioxide we pour into the atmosphere.

Aside from the real danger that the burning of coal poses for increasing the atmosphere's temperature, is this a source of energy that we should be willing to live with? The mining of coal is a dangerous and dirty business. When the deposits are deep, shafts are dug and the miners descend into the bowels of the earth every day. Rooms are carved out of the seams, and the miners loosen the coal from the walls using pickaxes or explosives. These rooms sometimes collapse, imprisoning miners; sometimes the room is filled with methane gas which can—and does—explode. Although the frequency of such accidents have been reduced in recent years, many miners are sacrificed each year to provide us with coal. It has been reported that more than 100,000 miners have died in the twentieth century as a result of cave-ins or explosions. This number is still increasing. The breathing of coal dust over a lifetime leads to a dangerous disease called "Black Lung."

No matter how efficiently coal is burned, it leaves an ash of very fine particles. These are then blown into the atmosphere, causing upper respiratory infections or diseases.

When coal deposits are close to the surface, deep mining is unnecessary, but the surface or strip mining which is then used has its own set of problems. Vast quantities of earth are scooped up to expose the coal and then heavy digging machinery removes it. This creates a terrible blot on the landscape. Legislation passed in several states forces the operators to restore the landscape. The cost of doing so is easily borne but requires an ample supply of water to replant the vegetation. Unfortunately, much of the coal that can be strip mined lies in parts of the United States that are not only scenic but have a short supply of water. Frequently the reclamation can be accomplished by invading the aquifers of the region, the underground water supply that must be replenished by rain. Since rainfall in some of these regions is small and droughts not uncommon, this is a dangerous practice.

Coal presents us with a dilemma. It is the most plentiful fossil fuel. It is versatile, and the technologies for extracting it from the ground and using it are well developed. It is prudent, as a public policy, to try to limit the harm extracting and using it causes to the environment, and to try to make it last as long as possible. However, it remains a nonrenewable energy source that is not likely to last as long as we anticipate because there are tens of millions of Earth's inhabitants who are waiting to achieve the standard of living of the industrialized countries; they would have to become coal consumers for this to be realized. There is also the serious problem of the greenhouse effect, which no amount of sanitizing coal can eliminate. Perhaps the greatest concern we should have is that we have become utterly dependent on commodities, coal and oil, that are bound to be in short supply in the relatively near future.

There is possibly another reason our society drags its heels in finding substitutes for fossil fuels. It may have faith that science will come to the rescue in any crisis. At the end of the 18th century, Thomas Malthus predicted that we will eventually run out of food to feed the world's people. He reasoned that because Earth's population increases geometrically and the production of food increases arithmetically, we are bound to arrive at a time of severe food shortage, and indeed mass starvation. Science and technology have so increased the efficiency of food production that the crisis has not yet come to pass and may never do so. Many believe a similar fate awaits the dire predictions about the consequences of relying so heavily on fossil fuels. Earth and its resources are finite. If we persist in increasing our demands for food, Malthus may turn out to have been right after all, no matter how clever our scientists and engineers. And the Cassandras may also be right about our fate if we rely too much on fossil fuels. We seem to have come to the limit of increasing the efficiency of agricultural production. In recent years the production of grains has scarcely been adequate to feed Earth's growing population.

As we shall see there are alternate renewable sources of energy which approach coal in their cost but require some financial encouragement to enter the market place. A public policy to encourage these alternatives is even superior to one which strives to make coal less harmful.

CHAPTER 6

NATURAL GAS

Natural gas is a fossil fuel, which consists principally of methane. Methane's chemical structure is one atom of carbon surrounded by four atoms of hydrogen. (In Chapter 17, we shall discuss what a splendid fuel hydrogen is.) Methane is rich in hydrogen and poor in carbon, so it is an excellent fuel in that it releases the smallest amount of carbon dioxide per unit of energy delivered of any of the fossil fuels. Thus, using natural gas will contribute least to the greenhouse effect. The actual numbers are 30 percent less than oil and 43 percent less than coal. It has other advantages, producing less carbon monoxide, fewer nitrogen oxide compounds, and less sulfur. Moreover it can be delivered economically to the user without any noxious hydrocarbons, which can be removed by the supplier and sold as feed stock for plastics and other products.

Natural gas is one of the end products of the physical and chemical processes that have produced the oil and coal. It is the gas that creates the gusher when an oil well is opened. Before its value was appreciated, natural gas was flared or burned at the well site (and still is in some countries).

Although natural gas had been used for many centuries, interest in it as an energy source began only at the beginning of the 20th century. An important energy source requires a distribution system which was lacking for natural gas until the 1940s. Pipeline construction began in the United States then and continued into the 1960s. As a result, gas could be easily and cheaply delivered to many remote parts of the country. A similar phenomenon took place in Europe. Some countries, such as Japan, had not yet developed any indigenous sources and found it too expensive to build pipelines. They developed a trade with Indonesia in liquid natural gas.

By the 1970s, natural gas was a major energy fuel. It became so popular that shortages appeared in the market and its price rose. In the late 1970s and into the 1980s the production of oil in the United States stabilized. This was an important cause of the shortages because oil production had been its most important source.

Well-meaning legislators established price controls to protect the consumer. This exacerbated the shortages. The low, controlled price further stimulated demand and the same low price created a disincentive for exploring new wells. By 1984, oil company executives wrote off natural gas as an important future energy source. When the price controls slowly disappeared natural gas use went up by 30 percent within a decade and is still rising. Its use has increased in every country of the world.

The rise in the amount of proven reserves of natural gas came from eliminating flaring in new oil wells, from finding new sources of the gas, in coal seams for example, and by discovering pockets of methane very deep in Earth's crust. This last type of natural gas exists at a very high temperature. Fortunately, methane is stable and does not decompose readily even at elevated temperatures. These methane pockets have no coal or oil near them and their origin is somewhat mysterious. A possible hypothesis for its origin is that this methane is the end product of the digestion of hydrocarbons by bacteria that exist deep in the Earth. However, Professor Thomas Gold of Cornell University has postulated that this methane is part of the primeval stuff of the universe. He believes there might be almost an infinite supply. Not many scientists believe this but if it is true, the reserves are incalculable. In any case lots of natural gas exists deep within Earth. This source is difficult and expensive to find but cheap to exploit once it is found. We do not know for sure how much is buried there.

The known proven reserves of natural gas are estimated to be likely to last about 150 years at the present level of use. Of course, if it were to replace substantial amounts of the other fossil fuels now being used it would disappear much sooner. However, the low estimates of how much gas there is must be taken with a grain of salt. In the past, the estimated size of the resource has always grown as its use has expanded. There is good reason to believe that the gas reserves have always been underestimated. As we have noted, the deep wells are expensive to find but cheap to use so companies have no incentive to identify more than they will need in the next few years. This resource seems to be pretty widespread. We surmise this is a result of explorations carried out in developing countries. For example, China, which is topographically like the United States, has scarcely scratched the potential it probably has. And so it goes throughout the world.

Another possible store of natural gas is methane hydrate. There are huge deposits of it widely distributed but deep underground. Methane hydrate consists of single molecules of methane trapped within crystalline cages formed by frozen water molecules. It had been discovered in the early part of the 19th century but had not been studied by geologists because it was found under conditions of temperature and pressure which are difficult to explore. The discovery of sporadic pockets of methane hydrate a few decades ago led several researchers to estimate how much might be in the ground. Beginning in the 1990s, much more data have been discovered. These new data seem to confirm an estimate made by geologist Gordon MacDonald, Director of the Institute for Applied Systems in Laxenberg, Austria in the 1980s, that methane hydrate deposits probably constitute the largest store of carbon that we know of underground. Others have conjectured these deposits might exceed all of the known reserves of coal, oil, and conventional natural gas. This conjecture has recently been given added weight by a discovery of lodes of methane hydrate hundreds of meters below the ocean bed off the coast of North Carolina. Researchers found methane hydrate and free methane there in concentrations about ten times previous estimates.

If these estimates are correct, and we somehow learn how to mine methane hydrate, and we learn how to recover the methane in these crystals to be used as a usable fuel, we will have a great deal of difficulty saying farewell to all fossil fuels. For this to happen there is a technology yet to be developed to take advantage of this lode of natural gas. There will be capital investments to be made should it become feasible to mine the methane hydrate and recover the methane. It will then be necessary to decide whether to make such investments rather than invest in renewable technologies which do not cause the greenhouse effect.

In the early days natural gas was principally used for heating. The first major breakthrough for an alternate use was for aircraft jet engines. In these engines the gas is compressed, ignited, and allowed to expand in the presence of rotating blades which drive a turbine. This configuration of machinery mounted on the ground and connected to a generator of electricity is called a gas turbine. Because methane burns so efficiently it makes for a very efficient generator. After natural gas prices came down and turbine technology advanced, natural gas began to be used extensively. The conventional steam turbine which uses pressurized steam as a working fluid managed to attain an efficiency of 33 percent, two-thirds of the energy of the fuel being lost. The gas turbine alone can attain efficiencies of 39 percent. If the excess heat of the gas turbine is used to run a steam turbine, still more of the energy of the fuel is usefully employed. The efficiencies for these combined cycle plants have been as high as 50 percent. And they are relatively inexpensive to build. One might even foresee individual buildings having their own source of electricity using combined cycle gas turbines.

Natural gas is very versatile. It should serve as a replacement for any fossil fuel in any application, including powering motor vehicles. Indeed there have long been automobiles and buses that have used liquid natural gas as their fuel. The only problem that has not been solved from an engineering point of view is the means of conveniently storing an adequate amount of fuel in the vehicle. It is unlikely liquid natural gas will replace gasoline any time soon. Nonetheless, many public utilities, municipalities and other government installations have used liquid natural gas powered vehicles over long periods. Recently the market for natural gas-driven vehicles is diminishing. However, the Japanese automobile manufacturer, Honda has announced it will start producing a line of automobiles that can use natural gas as a fuel. Whatever its ultimate fate as a motor vehicle, natural gas certainly could be considered as a backup should serious shortages of gasoline develop. What it really lacks is a distribution system similar to the ubiquitous gas station.

But there is an awakened interest in natural gas-driven cars as will be described in Chapter 15. During the transition to alternative energy sources, it would be important to develop natural gas as a fuel for aircraft transportation. Aircraft require a fuel with a high energy density because of the weight limitations inherent in flying machines. Alternatives to petroleum products for this use are very difficult to produce.

Natural gas is virtuous compared to the other fossil fuels, but it does have some defects. We have already noted that if handled carelessly, natural gas is likely to explode; this risk is possible to control. A nightmare is a port full of ships carrying liquid natural gas, where the proper precautions for storage have not been taken. The resultant explosion would be catastrophic. This has not happened so it is apparent that the risk can be kept in check.

Despite its virtues, natural gas is a fossil fuel and an uncontrollable contributor to greenhouse warming. Given a choice right now, the environmentalist would elect to use natural gas over any other fossil fuel. This points to a significant role for this gas. Use it in the transition until really renewable nonpolluting energy sources become common place.

CHAPTER 7

NUCLEAR ENERGY (FISSION)

Nuclear energy is an existing technology capable of supplying a significant amount of the energy now being generated by burning fossil fuels. It will not contribute to the greenhouse effect or spew the acids so destructive to the environment into the atmosphere. It accounts for nearly 20 percent of the energy used to generate electricity in the United States today, compared to the 70 percent that fossil fuels supply. In France, it is the other way around, with nuclear energy supplying 75 percent of that country's electricity.

To understand why the United States is so different, consider the problems that the generation of nuclear energy create. To understand these, we have to have some notion as to how this energy is produced. We have already described the structure of atomic nuclei; they consist of two types of particles: positively-charged protons and electrically-neutral neutrons, tightly bound together, each almost 2000 times as massive as an electron.

Different elements have different numbers of protons in the nucleus. The lightest nucleus, hydrogen, consists of a single proton. The next lightest stable nucleus, deuterium, heavy hydrogen, consists of one proton and one neutron. Going up the scale, the normal helium nucleus (sometimes called an alpha particle) has two protons and two neutrons. Still ascending, normal lithium has three protons and three neutrons. The heaviest naturally occurring nucleus, uranium, consists of 92 protons and 146 neutrons. This nucleus is designated as uranium 238, the number 238 referring to the total number of particles in the nucleus. Nuclei of elements sometimes can occur in nature with the same number of protons but different numbers of neutrons. These are called isotopes. They have the same chemical properties (chemical properties depend on the number of electrons in a neutral atom; since different isotopes have the same number of protons in the nucleus they have the same number of electrons to neutralize them electrically) but isotopic nuclei behave differently when their nuclear properties are involved. There is a stable isotope of the heaviest uranium nucleus that occurs in nature with 92 protons and 143 neutrons, uranium 235. Its abundance is 0.7 percent: one uranium 235 for every 140 uranium 238 nuclei. Uranium 235 has a stellar role to play in the generation of nuclear energy.

The basic research leading to the possibility of generating energy using the properties of the nuclei was motivated by the usual scientific drive to understand nature. Otto Hahn and Fritz Strassman discovered the process that liberated vast quantities of energy. It involved the breaking up of the uranium nucleus, a process that is called fission in analogy with the splitting of living cells. Physicists knew that much energy would be released as a result of this fission. Despite these discoveries practical utilization of this knowledge may not have occurred for generating nuclear energy had the United States not been involved in World War II and thus committed itself to investing billions of dollars to create a nuclear bomb. Harnessing the fission process to generate energy safely helped reveal the necessary ingredient for making the nuclear bomb. The engineering task for doing this was so difficult and the investment required so large, that it is unlikely that it would have been accomplished, at least not as quickly as it has, had not the government undertaken it with a different end result in mind.

Early in the 20th century the atom's nucleus was discovered. Once its size and constituents were identified, physicists became curious about the properties of nuclei and their possible reactions. Early on, experiments were conducted to strike nuclei with other particles. The object of these experiments was to disrupt the nuclei and make them break apart or transmute—and thus convert one element into another. Perhaps one motive for these experiments was to fulfill the alchemist's dream of changing lead into gold, although physicists never came close to this transmutation. The most important reason was to learn the properties of the nuclear force by observing the results of the breakup.

The neutron was one favorite particle used for bombarding nuclei. It is electrically uncharged, thus it is not acted on by the electrical positive charge of the nucleus and can penetrate it. Once inside the neutron collides with the nuclear constituents, shaking the nucleus up. The nucleus sheds excess energy achieved by the neutron by emitting a particle (or particles) or by emitting X-rays. The experimental data that the breakup furnishes has enabled physicists to understand a great deal about the structure of nuclei. The resultant nucleus is often unstable and will undergo additional changes. Such nuclei are called radioactive.

Radioactive nuclei always decay at a particular rate. The rate of decay is measured by a quantity called its half life. This is the time it takes for half of the radioactive nuclei in a sample to decay. Half of the nuclei are transmuted in one half life, a fourth in two half lives, an eighth in three half lives, and so forth (Fig. 7.1). Half lives can vary from fractions of a second to billions of years. A particular half life is characteristic of a particular radioactive nucleus and can serve as a signature for its presence. Nothing can change the half life of a nucleus—not any chemical or physical process. The uranium nucleus is naturally radioactive with a half life of 4.5 billion years which is almost the age of Earth. Nuclei heavier than uranium, the so-called transuranic nuclei, are all radioactive and have shorter half lives. This is the reason that uranium has survived as the heaviest nucleus. The others have disappeared by radioactive decay.

When most nuclei were bombarded with neutrons, the collision usually resulted in radioactive nuclei, which differed from the struck nuclei by having one particle (neutron or proton) more or one particle fewer. On rarer occasions the new nucleus might have two particles more or two particles fewer than the one bombarded. When Otto Hahn and Fritz Strassman bombarded the uranium nucleus they had a great deal of difficulty identifying the resultant radioactive products. The technique for making the identification is to find

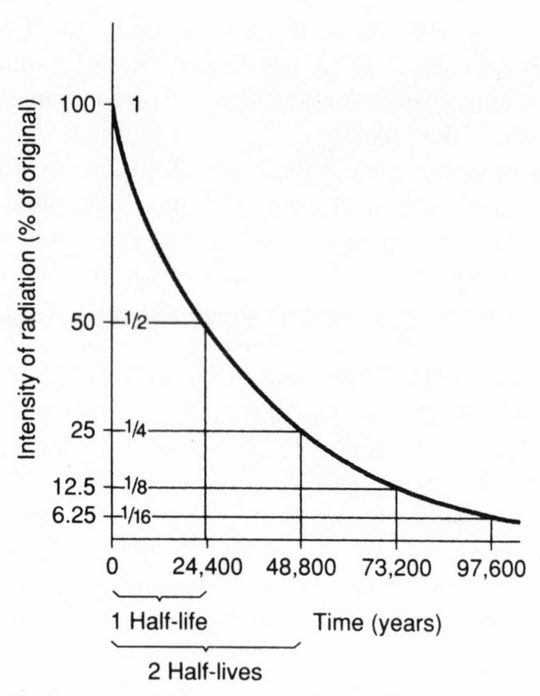

Figure 7.1. Rate of decay of a typical radioactive nucleus. (Reprinted with permission from the American Chemical Society.)

some chemical which would combine with a radioactive nucleus with a particular half life, precipitate out the combination and then remove the combining chemical. They tried using barium with one of the radioactive end products of the bombardment of uranium with neutrons. Barium did precipitate out the radioactive nucleus, whose identity they were seeking. However, the radioactive nucleus could not be separated from the barium. The rational conclusion would be that the radioactive nucleus itself was barium. Barium has 56 protons in its nucleus, uranium 92; it is hardly a nearby neighbor chemically. The idea that the resultant nucleus was barium was so far-fetched that experimenters continued to feel their experimental technique was at fault. Lise Meitner and her nephew Otto Frisch found a plausible explanation of how such a reaction might take place.

Some years before these experiments, Niels Bohr, one of the foremost physicists of the 20th century, pointed out that the forces holding nuclei together were similar to the forces holding the molecules in a liquid together. Many nuclear reactions could be understood on the basis of this liquid drop model. Meitner and Frisch guessed that a neutron striking a uranium nucleus might cause the liquid drop to oscillate, then elongate (Fig. 7.2) and perhaps develop a constriction at some point. Since the nuclear attractive forces operate only at short range, the repulsive electrical forces at the constriction could overcome the attractive nuclear force and cause a part of the nucleus to break away. According to this picture the part that broke away in this experiment was the barium nucleus. The remainder must be krypton with 36 protons because no protons can be lost in the reaction.

This picture of the breaking apart of uranium, soon named nuclear fission, was quickly verified. What made the explanation even more appealing was the discovery that barium and krypton were not the only ways that uranium underwent fission but the constriction could be formed at other locations and strontium and xenon and other pairs of elements could be the end products. (Otto

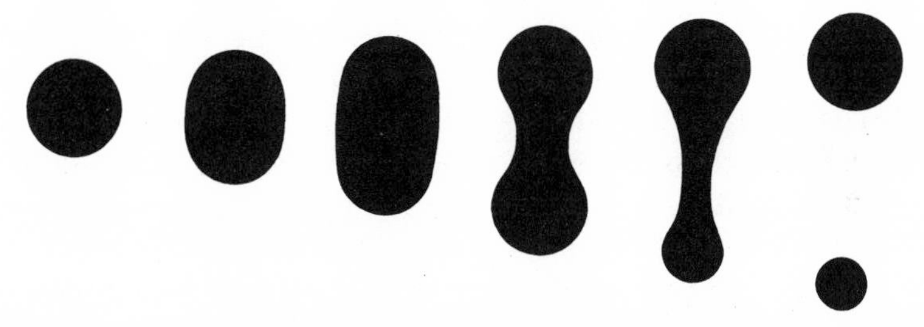

Figure 7.2. A neutron striking a nucleus can cause it to vibrate. The vibrations might cause the nucleus to deform until a narrow region is formed where the surface forces, trying to separate the nucleus, can overcome the attractive forces that hold the nucleus together. (Reprinted with permission from McGraw-Hill.)

Hahn received the Nobel Prize for his contribution to the discovery of nuclear fission. It is incomprehensible as to why neither Lise Meitner nor Otto Frisch shared this prize or even received one for their own discovery.)

Meitner and Frisch realized a great deal of energy would be liberated by this reaction. They were even able to estimate how much. Relatively early in the twentieth century, physicists had determined the binding energy of all of the known isotopes of the elements. Binding energies are the energies nuclei need to keep them together. When nuclear particles combine to form a new nucleus the protons and neutrons give up some of their mass, which is converted to the binding energy that holds the nuclei together according to Einstein's formula $E = mc^2$. The binding energy per nucleon of nuclei increases when going from lightest to heaviest up until iron. Beyond iron, the binding energies per nucleon become weaker until we reach uranium. A plot of this data as a function of the mass of nuclei is called the Curve of Binding Energy (Fig. 7.3). One can infer from this curve that if uranium were to split up we would end up closer to the maximum of the Curve of Binding Energy. The shape of the curve guarantees that uranium would be giving up some of its mass when it would undergo fission and this mass loss could be converted to energy. Meitner and Frisch knew the numerical details of the Curve of Binding Energy. They estimated that the energy released would be 20 million times the energy released in a typical chemical reaction.

Although the nuclear bomb is not the subject of this book, we can use the fact of its destructiveness to realize the enormous amount of energy released when mass is converted to energy. In the World War II bomb which devastated Hiroshima, Japan, only six kilograms of uranium were so converted. However, the bomb as an explosive device was inefficient; it had to contain 60 kilograms of fissionable material in order that six kilograms be available for conversion.

If we were to go up the curve from the light end and convert hydrogen to helium, for example, a process which provides us with the energy from the sun, we would also liberate a great deal of energy. This will be important when we discuss the energy we might be able to generate by nuclear fusion, the process of combining

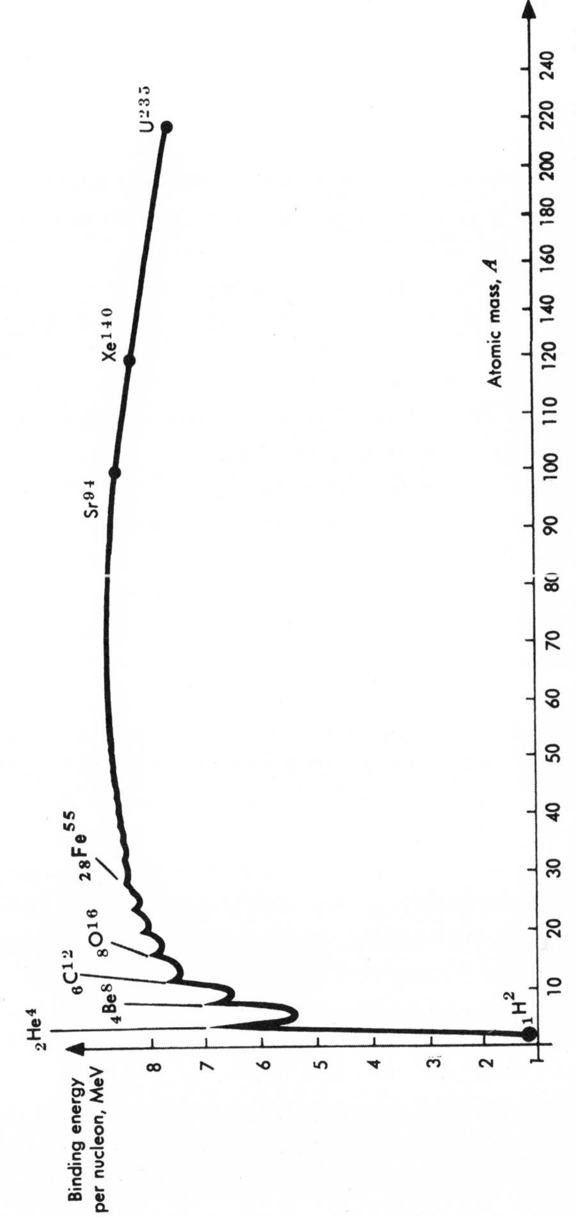

Figure 7.3. *The shape of the Curve of Binding Energy illustrates that when light nuclei combine and form heavier nuclei, an excess of energy becomes available. When nuclei heavier than iron break up and form lighter nuclei, excess energy becomes available. (Reprinted with permission from McGraw-Hill.)*

lighter nuclei to form heavier ones, to be discussed in the next chapter.

The discovery of the fission process did not lead immediately to either the possibility of constructing a bomb or the possibility of generating useful energy. Several additional discoveries had to be made and required a great deal of labor before either of these missions could be fulfilled. The first was the discovery that uranium 235, the lighter isotope of uranium, was responsible for most of the fission. This created a difficult engineering problem if one wanted to isolate uranium 235. In addition to the problem of the scarcity of uranium 235 in natural uranium, separating two isotopes with such a small fractional difference in mass (the ratio is 235:238) is not a trivial task. It was originally accomplished by a gaseous diffusion process which depends on the fact that the lighter isotope would, on the average, move more quickly than the heavier one. The method was successful but it required acres of equipment to isolate an appreciable quantity of the lighter isotope. There was a second method which used an electromagnetic principle to separate the isotopes. This method also yielded small quantities of end product and also required large expensive facilities.

The second problem was to discover if the fission reaction could sustain itself. For that to happen, because the fission is initiated by neutron bombardment, neutrons must be generated during the fission in sufficient number to keep the reaction going. With some difficulty it was determined that on the average 2.4 neutrons are released in each fission process, more than enough to continue it. It was also discovered that slow neutrons are much more effective than fast ones in causing the uranium to split.

The elements of a nuclear energy generator, called a fission nuclear reactor, should now come into focus. A sufficient density of the light isotope of uranium is required so that after a fission the neutrons created by that process could encounter other uranium 235 atoms with high probability to continue the reactions. Further, some substance (called a moderator) in the reactor is needed to slow down the neutrons, to do the fissioning efficiently. Starting with a small number of neutrons, each encounter produces two or more, each of

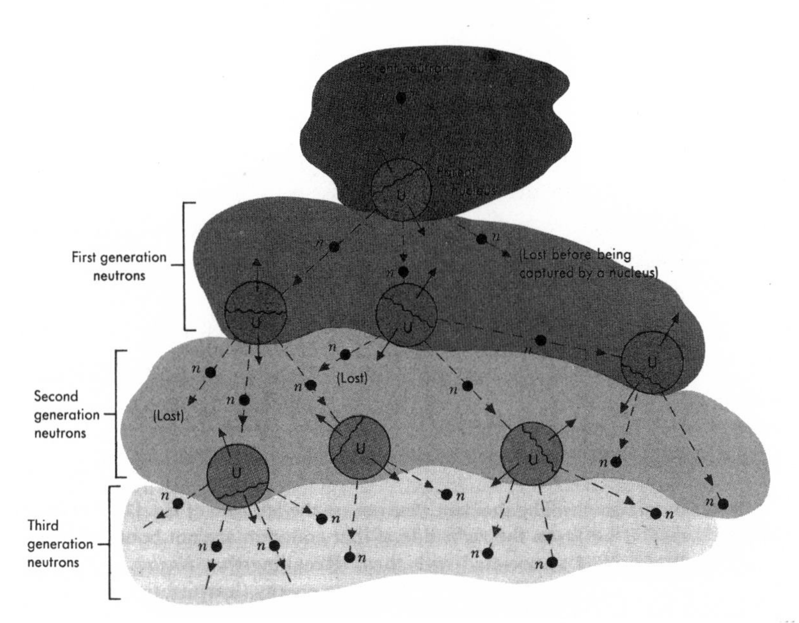

Figure 7.4. If at least two neutrons are produced per fission, the neutrons will multiply quickly after many collisions. In a real reactor, an element efficient in absorbing neutrons is used to limit the multiplication after the desired energy production level is reached. (Reprinted with permission from McGraw-Hill.)

these multiplication processes, generating energy (Fig. 7.4). For a reactor to operate without "running away" as a result of the multiplication, we must insert some control rods to absorb neutrons so that the energy production should be produced at a constant rate. Cadmium is the element of choice for this, although boron has similar properties. Finally, something is needed to absorb the energy created, which then can be converted to electrical energy. The substance most frequently used is water. Water can also serve as a moderator. It is important to use a substance as a moderator which is light; the hydrogen in water serves that purpose. A very heavy molecule would not slow down the neutrons. They would bounce off with velocities comparable to those they had before the collision. Water has the disadvantage of absorbing neutrons. In the United

States reactors, water is used as a moderator. To compensate for the neutrons absorbed 3 percent pure uranium 235 is admixed with the natural uranium. Canada uses heavy water, with the hydrogen replaced by deuterium, because it does not absorb neutrons. In these reactors the fuel does not have to be enriched with uranium 235. The reactor can be shut down completely by inserting rods of cadmium (into the uranium core) to absorb neutrons.

There is an unstated fear in people's minds when they think about nuclear reactors. It is that somehow the reactor could explode as a nuclear bomb. The principle of the bomb is based on the fission of uranium 235 and 238 by fast neutrons. The energy released by the bomb has to be released quickly—all at once. A bomb that depended on the fission using slowed down neutrons would never explode because of the time delay involved in slowing the neutrons down. Uranium 235 can undergo fission by collision with fast neutrons, albeit much less efficiently than by the slow neutron process. In designing a bomb one relies on the inefficient fission of uranium 235 and even less efficient fission of uranium 238 by fast neutrons. The more fissions that take place in a unit of time, the more effective the bomb. Nuclear bombs must have a very high density of uranium 235; nuclear bombs have close to 100 percent uranium 235. A reactor which normally has 3 percent of this isotope could not possibly become a nuclear bomb.

This does not exclude the possibility of other types of accidents occurring with reactors. Two most serious ones have occurred: Three Mile Island in Pennsylvania, USA and Chernobyl, Ukraine of the former Soviet Union. There was not a nuclear explosion in either of these cases. At Three Mile Island an equipment failure was the problem; a valve not closing was improperly interpreted by the operator. Cool circulating water, which normally prevents the reactor from overheating, was shut off and the reactor partially melted down. However, because the reactor was encased in concrete its design prevented any harmful radioactive material from being released into the atmosphere. In Chernobyl an unauthorized test was conducted with the reactor at low power. All of the safety devices had been shut off. The reactor became unstable. It overheated, and carbon, which was the moderator in this reactor, caught fire. Large amounts of harmful radioactive products were released into the atmosphere to be carried by the wind to all parts of Europe. The results were very serious not only for the people in the Ukraine, but for all of Europe. A commission investigating the results of the accident estimated there will be 70,000 additional deaths from cancer or radiation-induced disease in the next 70 years as a result of this accident. More recent estimates put the additional deaths at closer to 50,000.

Every accident, whether it is an airplane crash or a reactor mishap, produces improvements in the safety procedures associated with the technology. A president's commission investigating the Three Mile Island accident recommended design changes, but more important, it strongly recommended better training for the operating personnel. These recommendations have largely been implemented. Reactors like the ones in Chernobyl had been known to be of an obsolete design and no similar ones will ever be built again. The accident in Chernobyl was very serious, the one in Three Mile Island potentially serious. Both should be put into some perspective. There have been hundreds of reactors operating worldwide for scores of years with excellent safety records. None has had any accidents remotely as serious as either of the ones that have been cited.

There is a downside to using nuclear reactors as a source of energy that no amount of operator training can overcome. The fission reaction not only splits the uranium and produces fabulous amounts of energy but also produces as waste end products which are harmful to anyone exposed to them. We have described the fission as producing a variety of lighter elements such as barium, krypton, iodine, strontium, cesium, and many others. All of these end products are radioactive; they emit electrically-charged particles and gamma rays (which are a very harmful form of X-rays). These emanations can cause severe burns and induce all sorts of cancers in animals and humans. Some half lives of these radiations are very long—tens of thousands of years. As previously noted, nothing chemical or physical can be done to alter the half life of a radioactive substance. To protect oneself from the harmful effects of the radiation one must shield the radiator.

This, then, is the difficulty in using nuclear energy. Some of the harmful waste products can enter the food chain and must be shielded for many years before they can become safe to approach. This is especially true of strontium which, when ingested, replaces calcium in the bones and continues to be radioactive in the body, destroying the bone. The half life of radioactive strontium is about 30 years and must be shielded for several times 30 years before it is no longer dangerous. The radioactive end products of a fission are inevitably produced in a nuclear reactor and must be stored so that the environment is shielded from their emanations.

At the moment the wastes are being stored "temporarily" in water mixed with boron or are being buried in containers in places where the radioactivity will not harm the underground water supply. Plans are afoot to create a repository in the Yucca Mountains of Nevada. The repository must be very large, not subject to earthquakes, and far from any underground water supply. Many states have passed legislation saying "NIMBY—Not In My Back Yard." The Yucca Mountain repository is being subjected to all sorts of geological tests and will not be ready until the 21st century at the earliest. At the moment the wastes are stored in drums, which are immersed in pools of water mixed with boron to absorb some of the neutrons emitted. The public is so sensitive to the presence of radioactive wastes that objections will likely arise before the Yucca Mountain facility could be put into use.

Active research is being pursued to find a glass which would not disintegrate over a long period of time into which the radioactive end products of fission could be dissolved, providing a shield for the noxious emanations. Progress has been made but the estimates of the time such a glass will last before it disintegrates while carrying radioactive elements is not likely to be accepted easily by the public.

The problem is serious for the energy industry, as well as the country. Although the United States arms program has produced in excess of 300,000 cubic meters of radioactive wastes, the energy industry has produced fewer than 5000. The wastes of the energy programs are much more radioactive than those of the military ones. Thus the environmental hazards of both are comparable even though the energy programs produce fewer wastes.

Uranium, which is the heart of the nuclear reactor, is a nonrenewable source. The best present estimate for how long the supply might last is something over 100 years. Early in the development of the nuclear bomb, scientists found an alternative fissionable material which could expand the availability of uranium by a factor of by at least 50. When uranium 238 is bombarded with neutrons an end product is plutonium—a radioactive element that can last for tens of thousands of years and is itself fissionable when bombarded with fast neutrons. The bomb project viewed plutonium as a possible substitute for the light isotope of uranium, because it was so difficult to separate the fissionable uranium from the more abundant heavier isotope. After the first fission bomb was made, the ones that followed were constructed with plutonium.

A reactor constructed without a low mass moderator, so that the neutrons would not be slowed down as they are in an ordinary reactor, would convert the heavy isotope of uranium into the fissionable material, plutonium. The fissioning of plutonium would make lots of energy available but not all of the energy produced is needed to keep the reactor going. The excess energy would be available for other purposes. Thus a reactor of this sort would use the more plentiful isotope of uranium as a fuel. It is properly called a Breeder Reactor because, by creating plutonium, it makes more energy available for use than it consumes for its operation.

Since almost 100 percent of natural uranium is of the heavy variety, the Breeder Reactor effectively increases the uranium fuel available for energy production by at least a factor of 50. This is not the only possible source of plutonium. In an ordinary fission reactor not all of the neutrons are slowed down. The fast ones interact with the heavy isotope of uranium and produce plutonium by the same process as described above, although not as much. The spent fuel rods of a fission reactor can be processed chemically to recover this plutonium.

The plutonium raises a problem because it can be used directly to make a nuclear bomb. According to Ted Taylor, Ph.D., one of the world's experts on nuclear bombs, information on making a bomb is now available in the unclassified literature. A bomb designed such that as little as bomb grade fissionable material as the lead in a pencil is converted can cause a building like the World Trade Center in New York City to collapse. One kilogram can cause considerable civic damage; one made with five kilograms can devastate a city. (The bomb, of course, would have to consist of much more additional plutonium to have these amounts of materials converted to destructive energy.) If this material were stored in large quantities all over the world, it would require complex security measures and inventory controls to insure that no thefts would occur, a system unlikely to be instituted effectively. Indeed in 1994, there was increased traffic in illicit plutonium, thefts from stores in the former Soviet Union. Many of these traffickers have been apprehended, but the thefts point to the danger.

The United States recognized the potential danger of free floating plutonium some time ago. For the past ten years it has banned the construction of Breeder Reactors and commercial plants for reprocessing plutonium from spent fuel rods of conventional reactors, hoping that the world would follow suit. This is not the way things seem to be developing. The former Secretary of the Department of Energy has stated that the world would be better off if all stocks of plutonium were destroyed. Her counterpart in Russia, however, felt that stocks of this material are an economic asset, that despite the security risks plutonium poses it should be saved because it will constitute an energy source for the future when fossil fuels are used up. Many other countries, Japan among them, agree with this point of view and intend to invest heavily in constructing reprocessing plants even though there is no profitable commercial market for this material at present or in the immediate future. Japan's policy seems to have stemmed from the fact that it hoped to become energy self-sufficient through the use of Breeder Reactors. An accident in its experimental prototype Breeder in February 1996 has put the entire program in jeopardy.

The United States, of course, has stocks of plutonium stored in connection with its arms program. Recently the United States divulged how large their stockpile is in order to encourage other countries to follow suit. The United States also disclosed that it has shipped, over the years, small quantities of reactor grade, not bomb grade, plutonium for experimental and other benign purposes to many countries. However, in order to increase its trade in plutonium, the United States is contemplating changing its policy to the extent of not insisting on knowing to what use the recipient countries are putting the plutonium.

The United States and Russia are now negotiating about methods to dispose of the stores of plutonium in the world. Two proposals have surfaced. One is to encase the plutonium in glass in a way to make it impossible to recover the plutonium, and then to bury the containers. The second proposal is to construct reactors which will use up the plutonium but not breed additional fuel. Russia seems to be favoring the latter method, the United States the former.

We can see why the United States has been backward in its exploitation of nuclear energy. The public's perception of the risks of such use has put a brake on its spread. People are influenced by the catastrophic consequences of a Chernobyl disaster or by potential disasters that can be imagined as a result of the Three Mile Island accident. The image of the mushroom clouds of a nuclear bomb explosion and the horror of the nuclear aftermath of Hiroshima and Nagasaki have a strong grip on popular opinion. This image has affected the attitude of a substantial portion of the population to radiation and to the word nuclear, not only in the United States but in many other countries. For example, an effective way of sterilizing food so it can keep for a long period of time is to radiate the food. Radiated food emits no radiation. The miliary has been using irradiated potatoes without harmful results for some time. The Federal Drug Administration has authorized the sale of radiated chickens and other foods. It recently allowed the sale of radiated beef for hamburgers to avert repetition of the catastrophe where children died from eating tainted hamburger meat in a fast food restaurant. More recently 25 million pounds of beef were recalled and destroyed because it contained E. coli. This bacteria could be easily and inexpensively eliminated by irradiating the beef during production. Yet there are signs in grocery, produce, and meat markets, proclaiming they do not sell irradiated food, as an inducement for shoppers to patronize these stores.

Another example of consumer resistance to anything "nuclear" is that a medical imaging device comparable to a CT scan had to be changed from "Nuclear Magnetic Resonance" (NMR) to "Magnetic Resonance Imaging" (MRI) to allay the fears of patients requiring such a scan. Here we have an example of a perceived risk that is nonexistent.

Public reaction to nuclear energy is so negative that the safest nuclear reactor ever built, the Shoreham Reactor in Long Island, at a cost of over $2 billion has been mothballed. It has been purchased by the State of New York to alleviate the burden of higher rates for potential users. Presumably, it will be converted to a fossil fuel facility. So off-putting has the idea of radiation become that although chickens sterilized by being irradiated have been allowed to be sold in the United States for about ten years, producers have not rushed to install the necessary equipment in their plants. Nor have consumers clamored for the product. It remains to be seen how the recent permission granted by the FDA for sterilizing beef by irradiation will be accepted by the public.

We are constantly taking risks when we cross the street, get into a motor car, or fly in an airplane. We assume these risks because we perceive the benefits outweigh them. For some reason we do not do the same accounting when we compare nuclear energy generation to fossil fuel use. There are risks involved in burning fossil fuels. As we have noted there are miners who die mining coal; there is the potential danger of greenhouse warming due to carbon dioxide generation; there is the pollution due to the generation of acids of sulfur and nitrogen. We must reckon on the general environmental degradation due to strip mining, the pollution due to oil spills, the difficulty of conducting foreign policy with oil-producing countries (consider the recent Gulf War with Iraq) and the necessity of subsidizing some of these countries. These very real risks ought to be weighed against the relatively small risks of catastrophic accidents with reactors and the problems of storing the radioactive wastes produced by the energy industry.

Such irrational reactions mean one cannot be sanguine about the prospects that fission reactors will become a preponderant energy source, at least in the United States. Public opinion has cast its vote and no more nuclear power plants are being built here. Foreign countries have been, up until recently, largely exempt from the strong reactions to nuclear power plants. But they are beginning to feel some opposition now. Fortunately other renewable alternatives exist for replacing fossil fuels that do not raise such strong emotional responses.

CHAPTER 8

NUCLEAR ENERGY (FUSION)

We have seen there is a proven technology ready to supply a substantial portion of our energy needs—nuclear fission. There are, as previously noted, objections to increasing our dependence on this source, some valid, some invalid. Fission reactors do produce long-lived dangerously radioactive byproducts that are difficult to store, and fission reactors do require active safety systems which could fail in operation. They also produce bomb material, plutonium, which requires stringent security measures to prevent this dangerous material from falling into unauthorized hands. The potential for accidents, despite the safety systems, does prey on people's imagination so the fission technology is not readily acceptable to many communities. Finally, the uranium used in fission reactors, while not in short supply, is not a renewable source and is not that plentiful, if we foreclose the possibility of converting it to plutonium.

There is a possibility, however, of producing nuclear energy without having any of these drawbacks. The sun produces its vast amount of energy by combining four nuclei of hydrogen to form a helium nucleus and two positrons, a process called fusion. The mass of the helium nucleus and the two positrons is less than the four hydrogen nuclei. The excess mass available in this reaction is converted into energy, according to the Einstein formula. We are ascending in this reaction the Curve of Binding Energy (Fig. 7.3). As already noted, this conversion is so generous in energy production that one gram of hydrogen converted to helium yields the equivalent energy of 20 tons of coal.

The previous chapter's discussion of the Curve of Binding Energy indicated that nature has supplied us with a distribution of masses among the elements so that when the lighter ones (i.e. lighter than iron) undergo fusion there is a mass differential available after the reaction to be converted into energy. When heavy elements such as uranium break up into lighter ones (the process of fission) there is a mass differential available to be converted into energy. This is the source of energy in the nuclear reactors discussed earlier.

Clearly if we could duplicate on Earth what goes on in the sun our energy problems would be solved forever. There is an infinite amount of hydrogen available for use. Our planet is mostly water, and every molecule of water contains two atoms of hydrogen. Moreover, the end products of the fusion of hydrogen into helium produces neither radioactive nor reactive end products so the problem of waste storage disappears. It is possible to design fusion reactors so that safety is assured as a result of the physical properties of the device rather than having to rely on installed safety systems. The trouble is that duplicating what goes on in the sun on Earth turns out to be a formidable engineering problem.

The difficulty in bringing the "sun down to Earth" has to do with the fact that hydrogen nuclei, protons, are electrically positively-charged particles. For the fusion reaction to occur, the protons must closely approach each other in order to fuse. But positively-charged particles repel each other with a force that increases inversely as the square of their distance part. In order to overcome this repulsion the hydrogen nuclei must rapidly approach each other. One can accomplish this by increasing the temperature of protons. This requires a temperature of between 100 and 200 million degrees. And this points up one of the stumbling blocks we face in dealing with this problem: The temperature of the interior of the sun is only 15 million degrees. What kind of a container is available to house the protons at this temperature for the amount of time necessary for fusion to take place?

The sun, of course, does not have this problem. It does not need a container. The gravitational attraction of its constituents, and the high temperature it has reached because of this gravitational attraction, keeps a high density of hydrogen nuclei at a high enough temperature for a long enough time.

A fusion reaction was produced on Earth when the hydrogen bomb was exploded. To achieve the desired temperature and sufficient density of protons for fusion to take place, the bomb was set off by first exploding a fission bomb inside of it. This not only produced an initial high temperature but caused all of the protons in the fuel to be compressed in a short period of time so that fusion could take place efficiently. Of course, if we wish to use fusion as a source of energy available for general use, we would have to fashion a less violent way for the reaction to take place.

It is an engineering difficulty to keep the protons at a high temperature for a sufficient length of time at a high enough density for enough fusion reactions to take place to generate a usable amount of energy. Doing this would duplicate the sun's feat on Earth but would require overcoming enormous engineering difficulties. Hydrogen atoms kept at a temperature of millions of degrees will have their electrons stripped from the protons, creating a gaslike condition of matter called plasma. Plasma-like gases will expand with increasing temperatures. If we try to squeeze the plasma of protons to increase its density, we increase its temperature at the same time, which serves to decrease its density because the plasma expands. A successful fusion machine should have an increased density of protons for a long enough period of time to produce the desired effect. The conditions have been quantified in what is known as the Lawson criterion: The plasma density (in particles per cubic centimeter) times the time of confinement (in seconds) must exceed 10^{14} seconds per centimeter cubed. (This number is 10 followed at 14 zeros.) One can imagine that this is a condition very difficult to attain, hence the engineering problems.

In addition to the density–time problem, we have, as we have indicated, a containment problem. It is difficult to create a bottle to contain the high temperature plasma. The plasma at the densities used does not have enough energy to melt or vaporize the bottle. When the plasma touches the sides of the container it cools and is unable to sustain a high enough temperature for fusion to occur. The containment is accomplished by creating an external magnetic field in a direction parallel to the long axis of the bottle. According to the laws of electricity and magnetism, the charged particles will spiral around the direction of this magnetic field and keep away from the walls (Fig. 8.1). A secondary current is induced in the plasma, further confining it.

The plasma acts as an electrical resistance. When a current passes through a resistance it heats up. Unfortunately the resistance of a plasma diminishes as its temperature is increased so there is a limit to how high a temperature it can reach by merely passing a current through it. Sufficiently high temperatures have been achieved by injecting atoms which have been heated outside of the chamber but by this time it should be clear that the laws of physics are conspiring to make attaining energy by fusion very difficult, thereby taxing the ingenuity of engineers.

The described method of containing plasma is fine—until it reaches the ends of the bottle. This problem has been addressed by connecting the two ends of the bottle to form a torus (Fig. 8.1). A fusion reactor using this configuration for its main chamber is called a Tokamak. It was invented in the former Soviet Union by Andrei Sakharov and Igor Tamm.

The Tokamak is not the only method of confinement. However, fusion research is extremely expensive and very slow. Because of the expense involved in investigating other alternatives, the Tokamak confinement seems to have been the method of choice in continuing fusion research.

This fusion research has been exclusively funded by central governments. While a successful outcome would be of tremendous importance for the world, its success is so problematic that no private company would be willing to assume the financial risks. Fusion research has been going on in the United States, Western Europe, the former Soviet Union (now in Russia), Canada, and Japan for almost

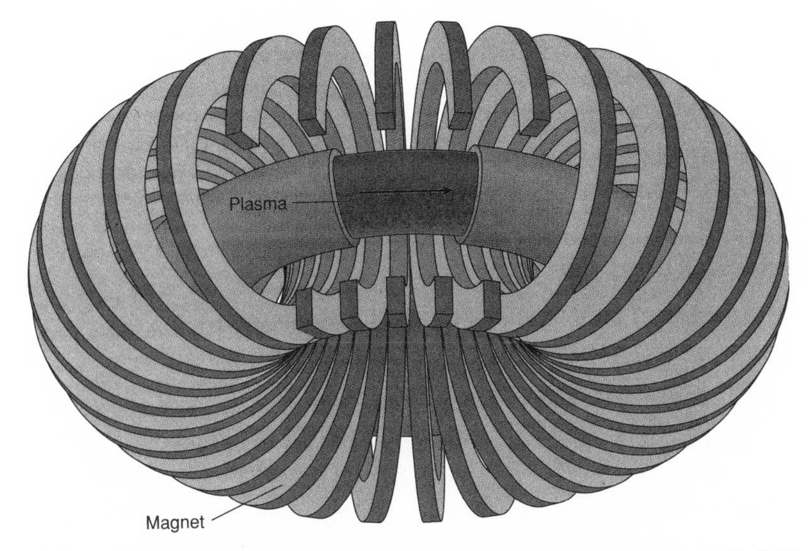

Figure 8.1. The plasma in a Tokamak is contained by introducing a magnetic field parallel to the walls of the container. The Tokamak is in the shape of a torus. (Reprinted with permission from American Chemical Society.)

40 years. The ever increasing costs have caused the countries involved to reduce their expenditures for fusion research. Still they have not entirely abandoned the project because progress has been made despite the lack of success in building a fusion reactor. The future would seem to call for some kind of international collaboration.

Some success has been achieved in trying to harness fusion energy in the Tokamak configuration. Two goals that have been set for success. The first is "break even." This means creating more energy as a result of the fusion than has been expended in initiating the reaction. "Break even" has almost been achieved but not by using ordinary hydrogen as a fuel. Rather it involves having two isotopes of hydrogen, deuterium and tritium, collide. As described, deuterium is a hydrogen nucleus consisting of one proton and one neutron. Tritium is a nucleus consisting of one proton and two neutrons. The reason these two isotopes were used is that they require less energy, and therefore a lower temperature to initiate the fusion. While deuterium is found in nature in a stable form and there is essentially an infinite amount that can easily and inexpensively be separated from ordinary hydrogen, tritium has to be manufactured using an isotope of lithium. The available supplies of this isotope are ample at the moment, but are limited. The expectation is that if fusion energy were commercially available, the energy produced could be used to separate this isotope from the ocean.

The apparatus that produced "Break even" is shown in Fig. 8.2. But this is far from having a successful source of fusion energy. The next step is "Ignition," which requires that enough excess energy be produced in the fusion reactor to keep the reactions going by supplying the energy to overcome all of the other losses in the device including the loss of energy that is being supplied to the consumer. If we consider how much energy is required to start the reaction, then the gap between Break even and Ignition seems very large indeed, and the machine necessary to achieve ignition is likely to be even larger and more complex than that shown in Fig. 8.2. Nothing has given us assurance that it could ever be achieved.

The United States Congress seems to have lost patience with the possibility of having fusion energy available sometime soon. Scientists in the fusion laboratory at Princeton, New Jersey have announced a breakthrough in eliminating leaks in the Tokamak that they expect might lead them much closer to "Break even." The Department of Energy has decided to close down the Tokamak project at Princeton. It is unlikely that the United States will be able to do the necessary development of a fusion reactor alone.

An international collaboration has been considering the feasibility of having all countries interested in a fusion reactor invest billions of dollars to construct a working device using available research results. The project is called ITER, for the International Thermonuclear Experimental Reactor. The practicality of this proposal has divided the scientific and technological community. Those for it think enough is known about the problems with this machine to anticipate a successful conclusion by 2008 for the project. They further believe

Figure 8.2. The Tokamak at Princeton that practically achieved Break even. Complicated as this installation seems, the ultimate fission reactor will be more complex. (Photo courtesy of Princeton University.)

that fusion energy could only become a reality if governments commit themselves, knowing that any stumbling blocks that arise will be solved as the project progresses. Those opposed are not as sanguine that enough knowledge about the difficulties are now known. They expect it might take a decade to complete this project, and if it fails, it would foreclose the possibility of a fusion energy reactor ever becoming a reality. Moreover, scientific and engineering progress in fusion research would be frozen during the decade ITER is being built. Cold water has been thrown on ITER by some knowledgeable scientists insisting that the confinement problem has to be reconsidered if a fusion reactor using only deuterium can ever be achieved. In addition, the international consortium has been reluctant to commit itself to the anticipated expense of the project. Further reservations have been expressed that the fusion reactor will not be totally nonpolluting because the apparatus itself would become dangerously radioactive. At the moment the expected starting date has been deferred for three years by the planning committee.

The fundamental fusion reaction does not produce any harmful radioactive products. It is not certain that there will be no radioactive pollution from a successful fusion reactor. Many neutrons are produced in the course of generating fusion energy and these, reacting with the walls of the device, might create radioactive products. It is almost certain that the components of the fusion device will become embrittled as a result of the neutron bombardment. These and other potential problems are on the minds of the planners. At the moment it seems as if those opposed to ITER will prevail.

A completely different research path is being investigated that might lead sooner to a successful fusion reactor. It is called inertial confinement. This research is trying to emulate more closely the way the hydrogen bomb has been made to explode. Setting off a fission bomb to initiate the reaction is out of the question. Instead small plastic capsules containing a mixture of tritium and deuterium are prepared. The capsules are presented in what will be the reactor so that they can be irradiated simultaneously from all directions by pulses of light coming from what is now expected to be almost two hundred lasers, each pulse lasting about three billionths of a second. The light striking the surface of the capsule heats it rapidly. An envelope of plasma is formed that vaporizes explosively and blows outward (Fig. 8.3). The outward explosion has a reaction on the fuel mixture by compressing it and increasing its temperature. The ultimate density reached by compressed fuels is estimated to exceed the density in the interior of stars, and its temperature is estimated to exceed 100 million degrees. With these temperatures and densities the fusion reaction can take place.

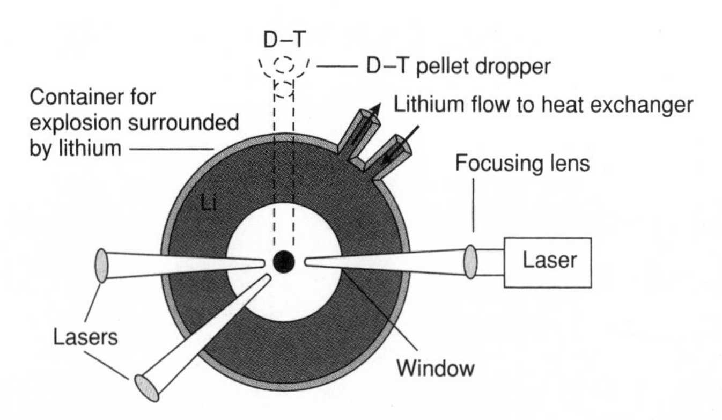

Figure 8.3. The basic process of inertial confinement fusion. A partial realization of this process is shown in Figure 8.4. (Reprinted with permission from the American Chemical Society.)

The United States Congress has already appropriated $191 million to be used for planning, preliminary construction, and operation of an experimental National Ignition Facility (NIF) to test the feasibility of obtaining fusion energy for general consumption based on this method of creating a fusion reaction. Ground has been broken at the National Laboratory at Livermore and construction has begun on a $2.2 billion facility.

Although the NIF is expected to create only tiny fusion reactions, the information obtained could lead to a nuclear fusion reactor. The facility will be very large, larger than a football field. Some of the technical problems of inertial confinement have been worked out on an experimental generator at the University of Rochester. Fig. 8.4 shows their facility and illustrates the scheme of using inertial confinement as an energy source. Even more extensive experiments have been done at the National Laboratory at Livermore. The NIF will be more complex and larger than either of these facilities.

Each capsule, filled with the hydrogen isotopes, is estimated to produce an energy of one-tenth of a ton of TNT, which is expected to be greater than the laser light energy input that precipitated the

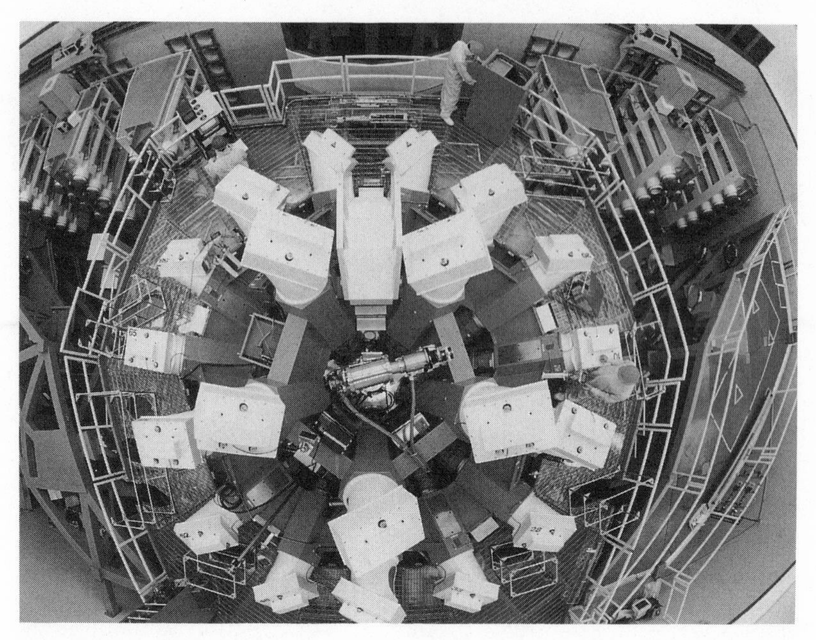

Figure 8.4. This installation at the University of Rochester has been used for preliminary testing of the inertial confinement principle. (Photo courtesy of the University of Rochester.)

reaction. For a practical fusion energy generator many capsules will have to be presented to the laser beams per second.

The NIF being planned is not expected to produce a practical device. (Perhaps it should have been called the National Breakeven facility.) Its mission is to demonstrate that it is feasible for inertia confinement to produce a generator whose output energy is sufficient to provide energy to the lasers, to provide energy to all of the other sinks of energy in the facility, and to be able to send substantial amounts out. If all goes well, the NIF will be completed in 2003 and its mission accomplished by 2006. A successful completion of the mission would have fusion-generated energy available a few years beyond that date.

The sudden interest of the United States Congress in fusion energy was prompted by the adoption of the Comprehensive Test Ban Treaty by the members of the United Nations General Assembly. This treaty says that "nuclear explosive testing will no longer be a permissible method of scientific inquiry." Should the United States abide by the provisions of this treaty it would limit its ability to refine further knowledge about how nuclear weapons work, and how they might be improved. The Department of Defense argues that the United States continues to rely on its nuclear deterrent, so that the safety, reliability, and improvements of its stockpile should not be compromised. The NIF's data would provide necessary information for conducting computer simulation modeling the performance and effectiveness of the nuclear weapons in the stockpile. When this was pointed out to the Congress by nuclear technicians in the Department of Defense, Congress responded. As a further incentive for a favorable congressional reaction, the NIF would assure a cadre of trained weapons' scientists and engineers, a necessary adjunct for being able to monitor and evaluate the effectiveness of the nuclear weapons stockpile.

Objections to the NIF have been raised by scientists, such as Hans Bethe and Frank von Hipple, because they believe inertial confinement experiments could ultimately lead to the construction of small hydrogen bombs and that by pursuing the NIF, the United States is in technical violation of the principle of the Comprehensive Test Ban Treaty. The courts have ruled that construction of the NIF does not violate the United States' compliance with this treaty.

Pure science would be an unintended beneficiary of these developments. The NIF should make available a state of matter at a density and temperature hitherto unavailable for experimentation. Such a development always opens fields of research whose consequences are unknown. Some in the scientific community are already salivating at the prospect of these new fields. An immediate application would be in astronomy because the conditions present in stellar interiors would be available on Earth, leading to a better understanding of stellar evolution and structure.

Should all the promise of the NIF as an energy source come to pass, the availability of fusion energy and the virtual infinite stores of hydrogen on Earth could indeed enable us to bid farewell to fossil fuels. We could also ponder the irony that one of the most serious threats to life on Earth, the hydrogen bomb, would become the lever for boosting us into an energy-rich age.

CHAPTER 9

DIRECT UTILIZATION OF SOLAR ENERGY

Burning fossil fuels or recently grown biomass, or harnessing the wind or running water are indirect ways of using the sun's energy. (The burning of fossil fuels is using energy taken from the sun millions of years ago.) It is possible to use the direct or the diffuse rays of the sun to supply the hot water needs of a home or to heat it. If one wished to use the sun's energy directly to generate electricity, it would be necessary to concentrate its rays on some working fluid of an engine and have the fluid research a very high temperature so the engine would operate efficiently.

One of the first written accounts of a high temperature achieved using the sun was related by the Roman historian, Galen. He described the burning of a Roman fleet by Archimedes in 212 B.C. by means of solar rays reflecting off mirrors. The story may be apocryphal but Archimedes did write a book entitled *On Burning Mirrors*.

Little more is recorded about the uses of solar energy for the next 20 centuries. By the 18th century, there is a record of a solar stove being invented for the cooking of food. This is an area of development that persists to this day. There is some hope that solar stoves could be introduced in parts of the world where the major accessible fuel is the wood of nearby trees. The denuding of an area of trees is a serious environmental danger that pollutes the air and degrades the quality of the soil. In addition, in the parts of the world where wood is the principal fuel, deforestation has already greatly reduced the availability of this fuel, producing a shortage of an essential resource.

Antoine Lavoisier, the founder of modern chemistry, experimented with solar furnaces. Using a lens approximately 50 inches in diameter, he coupled it to a smaller one and was able to achieve a temperature that could melt platinum—1760°C. The high temperature that can be achieved by using solar energy points to a possible application of this technology to burn the wastes of society, especially the toxic wastes. It turns out that concentrating sunlight is a very efficient method of vaporizing practically anything and reducing much of the toxic wastes to compounds that do not pollute. Moreover, the process simplifies the recovery of basic metals that might be part of these wastes. This is but one use for the high temperatures that can be developed by concentrating the sun's rays.

In 1875 August Mouchot made a notable advance in solar collector technology. The sun's rays were always concentrated to a point in an effort to achieve a high temperature. This is not the easiest way to heat a working fluid for an engine. Mouchot devised a collector that concentrated the sun's rays along a line rather than at a point. The most modern solar collectors for generating electricity use the same principle. Although they do not use Mouchot's technique, they utilize troughs, whose sides are parabolic mirrors (Fig. 9.1) that insure the rays will be concentrated along the axis of the troughs.

Following on the seminal work of Lavoisier in achieving high temperatures and the work of Mouchot, modern engineers have found it possible to use the sun's energy directly to drive an engine that in turn drives a turbine that drives a generator of electricity.

As we have seen, the higher the temperature of the working fluid the more efficient the engine. The surface temperature of the sun is 5500 degrees. Physical laws inform us that it is not possible to exceed this temperature no matter how much we concentrate the sun's rays. In practice it is possible to reach about 4000 degrees, but with some difficulty. It is easier to reach 1000 degrees but we need not go even that high to make a practical engine for producing electricity.

The sun's energy is cost free, so it should come as no surprise that attempts would be made to use this free resource to generate

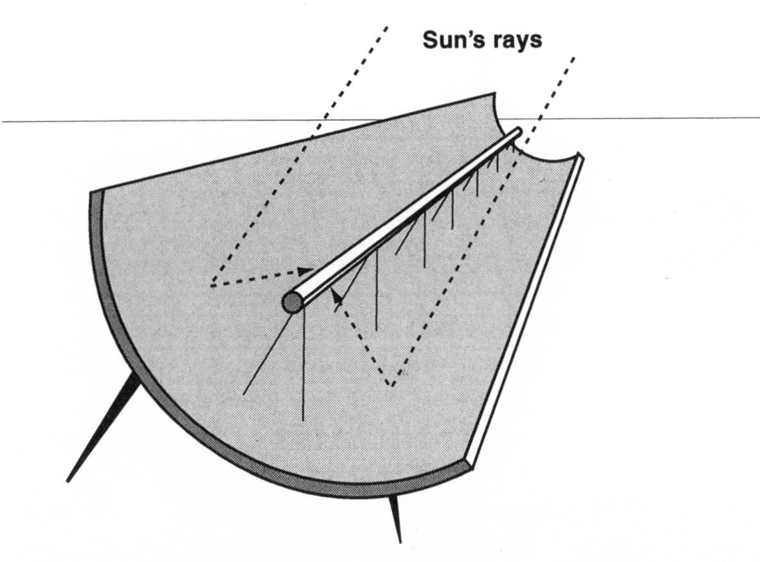

Figure 9.1. A parabolic trough concentrates the sun's rays along a line. The fluid to be heated could be water, oil, or even molten salt.

electricity directly. This has been done in several countries: the United States, France, Spain, Italy, Russia, and Israel. There are several viable possibilities for harnessing this energy to produce electricity directly. One can use troughs whose sides are the parabolic mirrors described above. Water or any suitable fluid circulating in the axis of the troughs would be heated.

A second technique is to use a collection of mirrors, called heliostats, to focus the sun's rays onto the top of a tall building. The working fluid of an engine that circulates around the region where the light is focused would be heated. The most efficient way to use the sun's rays directly to generate electricity is to have the engine as close as possible, preferably exactly in the region where the fluid is being heated. Positioning the engine elsewhere is inefficient because the fluid would cool off as it is circulated to a distant engine. Heliostats have been tried out experimentally. It is only parabolic mirrors that have been tested commercially.

Figure 9.2. The Luz installation in Southern California supplies electricity for 200,000 homes. (Photo courtesy of the Department of Energy.)

In 1984 an Israeli company, Luz International Ltd., installed a large electrical generation plant in Harper Valley, California (Fig. 9.2). The facility's design was based on an experimental installation operating in Israel. The plant was financed by a stock issue that was sold to the investment community. The distribution of the power was in the hands of the Southern California Edison Company.

This plant used the sun as its principal source of energy. The sun's rays were concentrated onto the focus of a parabolic trough of mirrors where a continuously circulating fluid was being heated.

This fluid was then used to drive an engine whose output drove a turbine and electrical generator. To get the maximum benefit of the sun's energy, a complex mechanical–optical system was installed to follow the sun's rays during the day so that the energy continued to fall on the working fluid as the sun's position changed. There was an auxiliary source of energy available for heating the fluid, natural gas, which was used when the sun could not supply enough energy to accommodate peak loads, or when the sun did not shine at all.

The costs of the initial installation was about $6000 per kilowatt and the costs of the electricity produced was 26.5 cents per kilowatt hour. Both costs were enormous compared to those of a standard generating plant, which are approximately $2000 per kilowatt for installation and 4–5 cents per kilowatt hour for generated electricity. (It should be noted, however, that especially in the northeastern part of the United States consumers can pay more than 10 cents per kilowatt hour for electric power.) Southern California Edison could make a profit selling this more expensive power only because the government was providing subsidies for the construction of facilities using renewable energy sources.

What followed on the initial installation were eight other Luz type plants, each increasingly more efficient. The last one cost about $3000 per kilowatt to install and produced electricity costing about 9 cents per kilowatt hour. This remarkable progress was obviously still not competitive with the costs of fossil fuel plants. In 1991, the government subsidy program was canceled as the price of fossil fuels declined. In the face of these events the company could not survive and declared bankruptcy. Just prior to going bankrupt the Luz organization had plans for a plant which would cost about $2000 per kilowatt to construct and would produce electricity costing about 8 cents per kilowatt hour. But absent the government subsidies, the stock issue for this plant did not find any takers and the bankruptcy ensued.

The nine plants built between 1984 and 1990 are still functioning, churning out power in the southern California desert. The rights to the technology were purchased by a group of Belgian investors who established a reorganized company in Israel called Solel. Presumably more research is being done. There are rumors that parabolic trough installations will be constructed in countries with lots of sun, lots of space, and lack of access to fossil fuels, such as Brazil, India, Morocco.

It may have been a mistake to abruptly abandon the Luz method of generating electricity. The size of the installation was enormous. It covered almost 2000 acres and was able to supply enough electricity for almost 200,000 homes. The Luz company had been able to decrease the cost of the electricity generated by a factor of three while it was in operation and additional developments leading to further cost reductions were underway. The government provides substantial subsidies to the fossil fuel industry through the depletion allowance. A more modest subsidy for such a promising renewable nonpolluting source might well have gone some distance to lessen our dependence on fossil fuels.

It may seem surprising that even though solar energy is free, it cannot successfully compete with fossil fuels. This indicates how unreasonably inexpensive fossil fuels are. More than that, a new complex technology requires an initial capital investment to compete with fossil fuel-generated energy whose initial investment has largely been written off. It was the capital cost of the initial equipment in the Luz scheme, the cost of its installation and maintenance, that caused its costs to exceed those of the fossil fuel providers. New technologies will either have to be so unrealistically inexpensive to start and to run or the marketing of new technologies will have to avoid direct confrontation with fossil fuel generators. Either of these alternatives is iffy. More likely the governments of the world will finally have to realize that it is in their interest to provide subsidies to promising emerging technologies. Fossil fuels will not last forever. As they become scarcer they become more expensive. Perhaps the alternative is to be patient, but it would be dangerous to be so patient as to bring us to the brink of catastrophe.

The Luz system, despite its promise, still would pose some difficulties other than financial ones. It requires large areas with lots of sunshine—not always available in sufficient quantity in highly industrialized countries. The sun does not shine all of the time so some storage for the energy generated has to be provided. Or some backup source has to be made available, as it was in the Luz system which provided natural gas for such a contingency. But the Luz system is so close to being able to compete that it, or some development building on its success, is very likely to play a role in supplying energy when there is an acute shortage of fossil fuels.

In the meantime, other research in this area is being carried out in Australia. David Mills of the University of Sydney has made extensive improvements on the Luz technology. He claims to have a design of a plant that could produce power for about 6 cents per kilowatt hour. Moreover his design does not require any natural gas supplement. By storing the sun's energy in a bed of hot rocks, his installation was able to run for eight hours in the absence of sunlight.

Another avenue of research is the heliostat, whose advantage over the parabolic trough is that it can achieve a higher temperature of the working fluid, thus making the engine running the turbine more efficient. The United States Department of Energy built an experimental model of a heliostat in California. In this early attempt to study its feasibility as a practical device, the sun's energy was collected atop a tower about 80 meters high with the rays being reflected from about 16 acres of mirrors around the tower. A working substance was pumped from the ground to the roof to be heated and then used to run an engine. The maximum temperature achieved was about 600°C. The cost of the installation per kilowatt of electricity produced was about $10,000—a very high cost. The installation cost was so high that the Department of Energy thought there was no future in pursuing this development and dropped the project. A second design is available should interest ever reemerge. This design would provide ten times the amount of energy with an estimated cost for the installation of about $3000 per kilowatt, comparable to the cost of a nuclear facility, with the expectation that the cost of the energy generated would be reasonable.

The principal drawback of the heliostat is that power is interrupted on a cloudy day or even when clouds pass by on a sunny day. The start-up time after an interruption was considerable in the early models. Recently, promising research has been done in separating

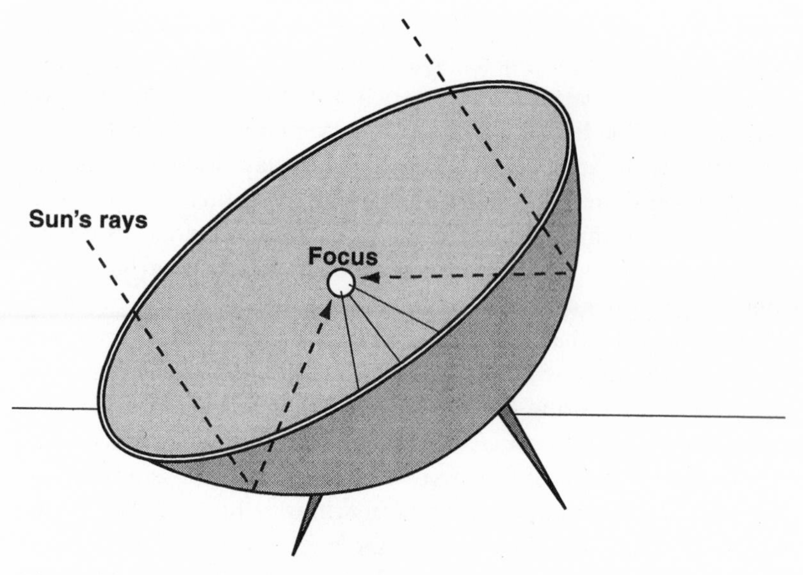

Figure 9.3. A parabolic dish to concentrate the sun's rays. A dish to use solar energy to generate electricity would be very large and its focus could accommodate an engine.

the turbine's operation from the solar energy received by using sodium or molten salt as a working substance which can store thermal energy. Yet there are no plans for any serious attempts to use this technology as a practical way to delivery electricity commercially.

The third method for directly using the sun to generate electricity is to use parabolic dishes (Fig. 9.3). By tracking the sun and having the rays come together at the focus of the parabola much higher temperatures can be achieved than either with parabolic troughs or heliostats, with consequent improvements in efficiency. In the first attempts to use this technology, the engine was at a distance from the parabolic reflector. As already noted, such an arrangement reduces the efficiency of the engine because the working substance cools off as it makes its way from the dish to the engine. This design was adopted because of the difficulty in constructing engines which could fit into the small region of the focus of a parabola. But recent developments of improvements in the so-called Stirling engine have made it possible to have small reliable self-contained engines at the focus. This is a technology whose considerable promise remains to be exploited.

A more successful strategy for competing with fossil fuels in the near future might be to eschew large centrally located sources of electrical energy. Instead the concept would be to try to use the direct and diffuse rays of the sun in installations that are small, widely distributed, and immediately economical, and involve devices which can be improved incrementally.

One model to imitate is the development in Israel of an industry that uses the sun's energy for hot water heating. Starting in 1940, 900,000 such heaters have been sold in this country of 3 million people, heating 83 percent of the hot water used there. Solar hot water heaters are used in the United States too, but our per capita use is much less. The system used for heating water is the flat plate collector. It is suitable incidentally not only for heating water, but also for heating an entire house.

The flat plate collector is a low tech device. It consists of a metal plate, painted black so that it can absorb the sun's rays effectively, with metal tubes (containing pumped circulating water) firmly attached to the plate (Fig. 9.4). The metal plate is put into a box. The back and sides of the box are insulated to prevent conductive heat loss. Two sheets of plate glass cover the box. They permit the sun's rays to enter the box. An air layer or vacuum between the plates cuts down on the convective heat loss. In addition, these plates also convert the collector into a greenhouse and prevent the escape of the heat. Temperatures of 140 to 160°F can be obtained even in cold weather, with this device.

The heated water can then be used directly or stored in an insulated tank to be used when there is no sun, such as at night or during cloudy days. A 1500 gallon insulated tank can retain its heat for four to six days.

If a flat plate collector is used to heat a building, it may not generate a warm enough temperature because the water cooling it

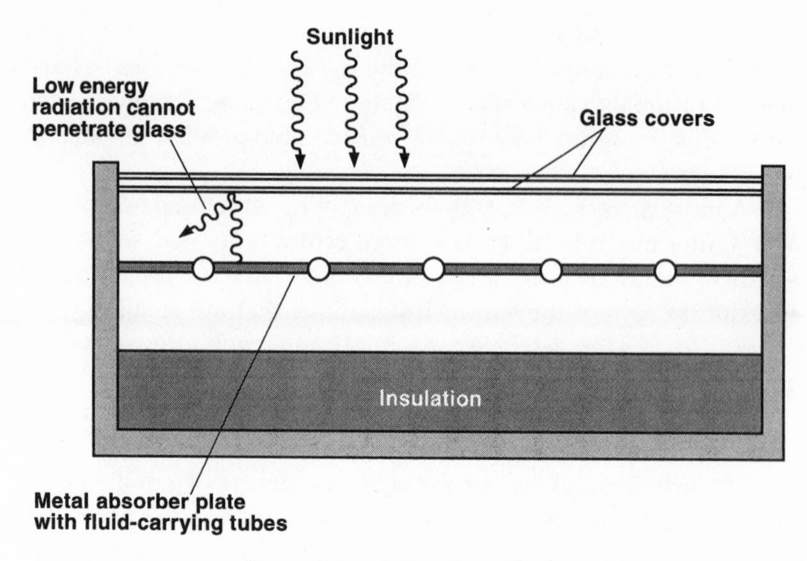

Figure 9.4. A flat plate collector.

must circulate throughout the building. The collector can be improved by using lenses to focus the sun's energy and by having the collector track the position of the sun. These improvements are very expensive additions to the system. Only when this method of heating a building is taken seriously will some less expensive development be invented.

There are more elaborate and better ways of storing solar heat in addition to using the insulated water tank. Lots of concrete and stone can be used in the building design or heat can be stored by warming a pile of rocks in the house. Some homes have auxiliary greenhouses and these greenhouses are used as a heat storage facility.

The sun's heat can also be adapted to cooling or even refrigerating. Refrigerators and cooling systems have been sold using gas as the source of energy. The same system using the sun's energy instead of gas has successfully been used for this purpose.

About 25 percent of the energy used in the United States is used for heating and cooling buildings. Virtually all of the applications involve the use of fossil fuels. Substituting the sun's energy in all buildings would go a long way in replacing fossil fuels with some renewable source of energy. Amory Lovins has been a pioneer in demonstrating that it is possible to cool and heat a home in a cold climate throughout the seasons without using anything but renewable sources of energy. But this tour de force comes at a price. For renewable sources to replace fossil fuels, there must be a relatively short payback time. This is the time needed to save an amount of money using present sources of energy equal to the installation costs plus the operating costs. Three or four years is good—ten years is an upper limit. This time depends on the complexity and cost of the system required, the climate of the installation site, current interest rates, and the cost of the energy source currently used. Except in sunny countries (such as Israel or the southern part of the United States) and for simple systems which only provide hot water, the payback time is frequently too long. During the oil crisis, the United States government provided tax credits to stimulate the use of flat plate collectors throughout the country. In the middle 1980s, the credits disappeared, and so did the market for these devices.

One cannot be too sanguine about the prospects that the alternative heating and cooling systems for buildings will cause significant conservation of fossil fuels any time soon. Installation of solar-powered heating or cooling devices can only be done for new buildings. Should these devices be installed in 90 percent of the new buildings it is unlikely they would reduce contributions from conventional sources by as much as 10 percent in ten years. Buildings have a long useful life and new construction is only a small fraction of the existing building stock. These savings would still be substantial from the point of view of the amount of pollution prevented and the amount of fossil fuels saved but these estimates indicate how far we still have to go.

One potential important application of the direct use of solar energy is to replace the use of firewood with a cookstove that uses the direct rays of the sun. This is not a significant problem in the United States, but it is an important one in other parts of the world. There are hundreds of millions of people in the developing world where cooking is their most important energy use. The widespread deforestation in those countries has caused great hardship and ecological damage. Fortunately, many of these countries have lots of sunshine and are good candidates for solar cookstoves.

Solar cookstoves have a history that date back to the middle of the 18th century. Attempts to develop solar cookers which would catch on have not succeeded until the last quarter of the 20th century. In 1976 an inexpensive stove which could be easily shipped and assembled was developed. This stove seems to be catching on especially in areas where wood fuel is in short supply. This is a case where direct use of solar energy will not replace fossil fuels but where it might have a desirable environmental effect.

CHAPTER 10

PHOTOVOLTAICS

Imagine a source of energy that is infinitely renewable, nonpolluting, lasts indefinitely, has a low maintenance and operating cost, and could, in principle, be able to satisfy almost all of the energy needs of the world. Such a source is represented by devices that convert solar energy directly into electrical energy. These are called photovoltaics.

Although the fundamental process of converting solar energy directly into electricity was discovered early in the 19th century, it remained only a laboratory curiosity until the 1950s. In 1954, the first practical solar cell was designed and made by Bell Laboratories. Still the cost of the electricity generated by the first primitive cells was so high that there was little incentive to develop the technology further. Its initial important use was in the NASA space vehicles where costs did not matter and access to electrical energy and reliability were the most important considerations.

It was not until the 1970s when oil prices shot up precipitously, generating acute concern about the future of fossil fuels in our economy that the industrialized world was forced to consider photovoltaics seriously as a terrestrial power source. To encourage the development of alternative sources of renewable energy the governments of the world and in particular the United States government invested in research in photovoltaic devices and provided subsidies for their manufacture and use.

One reason it was possible at that time to consider photovoltaics seriously was that the materials and the technology needed for the manufacture of the transistor—the device that has replaced vacuum tubes—were the same as those required to manufacture photovoltaic devices. Silicon, the basic material for making these devices is the second most plentiful element on Earth. The technology and techniques for manufacturing transistors were already well developed in the 1970s.

Electrical current is a moving stream of electrons. What induces them to move is a difference of potential or voltage between the region whence they originate and the region to which they are expected to go. Potential difference is the electrical equivalent of pressure. As the air moves as a result of the difference of pressure between two regions, electrical charges move because of a difference of potential. A battery can supply a difference of potential. A difference of potential exists in a region between a positive and negative charge. An electron placed between a positive and negative charge will move spontaneously toward the positive charge and constitute a current. Once an electrical current flows electrical energy can be extracted from it.

All matter consists of positive and negative electrical charges. A photovoltaic cell is designed to have a potential difference created in the cell. When light shines on the cell, it frees up electrons within which move because of this potential difference. A current is thus created and can be drawn from the cell to do work.

If sunlight is used as the light source, the cost of the energy is zero. These cells, which have no moving parts, are environmentally benign. If the cell itself is protected from the environment, it will last indefinitely. While each photovoltaic cell can produce only a small amount of energy, the cells are small and modular; they can be connected together to produce large amounts of energy. Photovoltaic cells can not only run small hand-held calculators but can be the basis of large electric utility installations.

The expense of generating electricity is the biggest obstacle in the way of more extensive use of energy from photovoltaic cells. Despite enormous progress made during the last decades in reducing this cost (it is now less than one two-thousandth of the cost of the power from the cell originally installed in the space vehicles), it is expensive compared to electricity generated using fossil fuels. It is still 30 cents per kilowatt hour compared to about 6 cents per kilowatt hour for energy generated using fossil fuels. However, the cost of the fossil fuel-generated electricity does not include the cost of pollution control. With this cost added and the prospects of reducing still the cost of the energy produced by photovoltaic cells, the differences will narrow.

Despite their high cost, photovoltaics have found a number of useful niches where it is cheaper to use them than other sources. Bell Laboratories experimented with these devices in the 1950s to develop a reliable source of electricity for telephone relay stations located in isolated areas far from conventional sources of electrical power. At the original cost, it was too expensive, but at the present price it is a welcome addition as a power source. In addition, photovoltaic cells have been used in remote lighthouses, microwave relay installations, roadside call boxes, and other installations in isolated areas. Recently, they have been found useful in vacation cabins and powering rural water pumps. They also provide a rural electrical power source where the cost of installing distribution lines (which run as high as $30,000 per kilometer to build) overshadows the cost of the electricity these devices produce.

The Japanese have developed inexpensive photovoltaic cells which produce currents small enough to power hand-held calculators and wrist watches. They have sold more than 100 million products using these inexpensive photovoltaics.

The greatest short term impact of photovoltaic devices might be in developing countries. There, 2 billion people lack electricity and depend on smelly, dim, kerosene lamps for their nighttime illumination. The cost of photovoltaic devices to replace the kerosene lamps (about $500 for a five-watt system, which includes wiring and a battery to store the electricity when the sun is not shining) is still beyond the resources of these people. But a number of private and public programs are how supplying funds for installing this electrical power, among them the World Bank and the United Nations. Larger installations are also being partially subsidized by the World Bank. Mexico, Sri Lanka, and other developing countries have had hundreds of thousands of photovoltaic panels installed on the roofs of their buildings. It is ironic that the basic energy needs of the world's poorest populations are beginning to be met by the most sophisticated energy technology yet developed.

It may seem odd that the cost of photovoltaic electricity is so high when the basic source of its energy is free. To understand this, we must explain how the photovoltaic cell miraculously converts light directly into an electrical current. After this explanation we can show what steps have already been taken and what steps will probably be taken in the future to reduce the price of photovoltaic electricity.

All materials can be classified according to their ability to carry a current: nonconductors, conductors, and semiconductors. Glass and hard rubber are in the first category; metals such as copper and aluminum in the second; and silicon, carbon, and germanium in the last. The electrical current-carrying facility of a substance depends, on a microscopic scale, on the number of electrons that can move freely within the substance. A nonconductor's electrons are securely bound to their constituent atoms and thus cannot move. Conductors have large numbers of electrons, which are free to move under the influence of even a small potential difference. Natural semiconductors have relatively few electrons capable of unrestrained motion and can sustain only small currents for a given potential difference.

Silicon is a semiconductor that has played—and still plays—an important role in photovoltaic technology. A crystal of silicon consists of many silicon atoms held together by mutual attraction in a regular crystalline configuration. The electrons of the silicon atoms, which are responsible for the mutual attraction of the atoms, are in very stable positions in the crystal. It requires some considerable energy to dislodge them. Any substance at a finite temperature has some of this constituent electrons in motion; its temperature is a measure of the energy of this motion. Only a few electrons are dislodged and are free to move in semiconductors at room temperature. These few are the ones that carry the silicon current. Photovoltaic material is created when: 1) the number of electrons free to move are substantially increased; and 2) a difference of potential is created inside the material so that these electrons can produce a useful current when light shines on the cell.

Atomic structure was accurately described in great detail by the theory of quantum mechanics during the first part of the 20th century. We now know that the electrons in an atom are arranged in a series of shells around a nucleus. Each new shell is at an average distance from the nucleus greater than the preceding one. There are four electrons in the outermost shell of silicon. These electrons are shared by neighboring atoms. In so doing, according to the insights provided by quantum mechanics, they hold the atoms of silicon together to form the crystal (Fig. 10.1). Since they are so rigidly attached to the atoms most cannot move except for the few that are dislodged due to their temperature-dependent motion.

Surprisingly, the first step in making a photovoltaic cell out of crystalline silicon is to add an impurity. This process is called doping. Arsenic has a single electron in its outermost shell. If arsenic is added to the silicon, these atoms will take their place in the crystal structure (Fig. 10.2). But the outermost electron of an arsenic atom will neither contribute to the "glue" holding the silicon atoms together nor will it be tightly bound to its atom. Under the action of even a small force the electrons will be able to move easily through the crystal to create a current. The combination of a silicon crystal with an arsenic impurity is called an "n" type semiconductor, "n" for the extra mobile negatively-charged electron. If other impurities, such as phosphorus or antimony, were added these would produce approximately the same effect as arsenic.

There is another type of semiconductor that can be made by adding a different impurity. If, for example, gallium were added to crystalline silicon, the resultant semiconductor would have different electrical properties. Gallium's outermost shell of electrons is incomplete. It would, given the opportunity, like to add an electron to it. Thus when added to silicon, gallium pulls an electron from a neighboring silicon atom and leaves a positively-charged "hole" in a silicon atom (Fig. 10.3). If a voltage were to be applied to the material, successive positive holes would be filled by electrons from

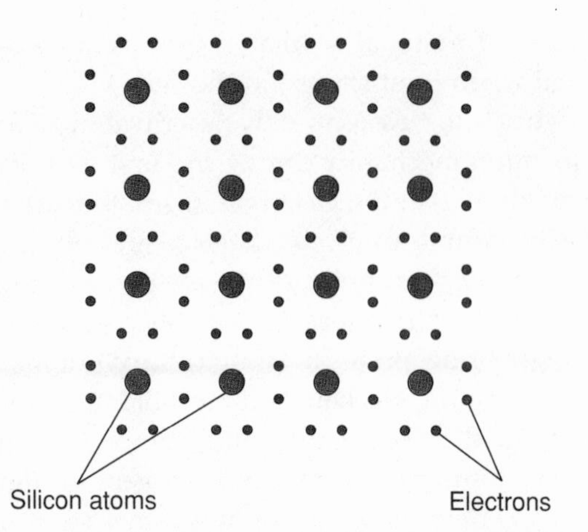

Figure 10.1. Silicon atoms in a crystal. Eight electrons are shared among neighboring atoms. This sharing, according to quantum mechanics, provides a force which holds the crystal together. The bond is called a covalent bond. (Reprinted with permission from the American Chemical Society.)

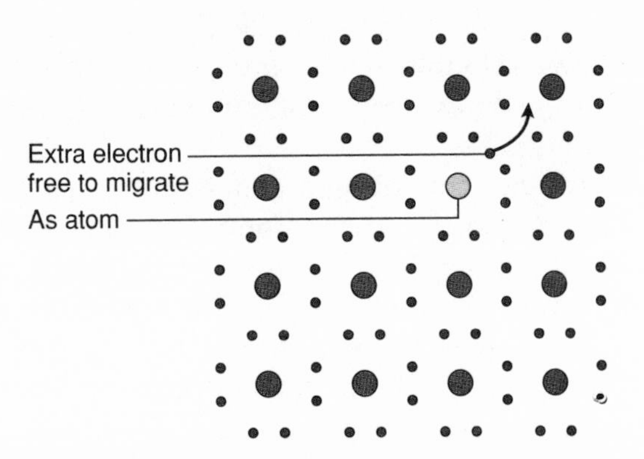

Figure 10.2. Arsenic when used to dope the silicon, provides an extra electron, which can migrate if a potential difference exists within the crystal. (Reprinted with permission from the American Chemical Society.)

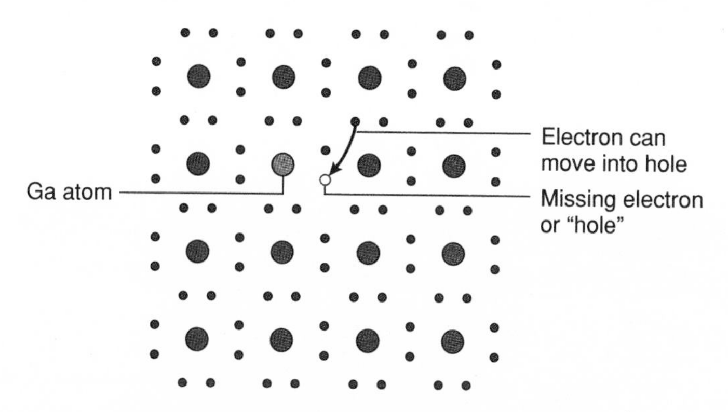

Figure 10.3. When added to a silicon crystal, gallium attracts electrons from neighboring atoms to its incomplete shell, creating a positive hole. If a potential difference exists within the crystal this hole will be propagated through the crystal, generating, in effect, a current of positive charge. (Reprinted with permission from the American Chemical Society.)

successive silicon atoms. The progression of holes is equivalent to the motion of a positive charge and is opposite to the motion of the electrons. A silicon crystal doped with gallium is called a "p" type semiconductor, "p" for the synthetic positive current produced by the succession of holes. Silicon doped with other elements, such as boron or indium, would produce approximately the same effect.

Interesting things happen when n-type and p-type material create a junction (Fig. 10.4). Some of the excess electrons in the n-type material cross the junction boundary because they are attracted to the positive holes. Some of the holes migrate across the boundary to the n-type material, another example of opposites attracting. The separation of the charges creates the voltage difference we are seeking, across the junction. If we now dislodge some of the electrons from their stable positions in the crystal by adding the energy of light shining on it, the liberated electrons will move under the influence of this potential difference to create a current. This current can be brought outside of the cell by externally connecting the n-type semiconductor to the p-type with a wire and be tapped to do some work.

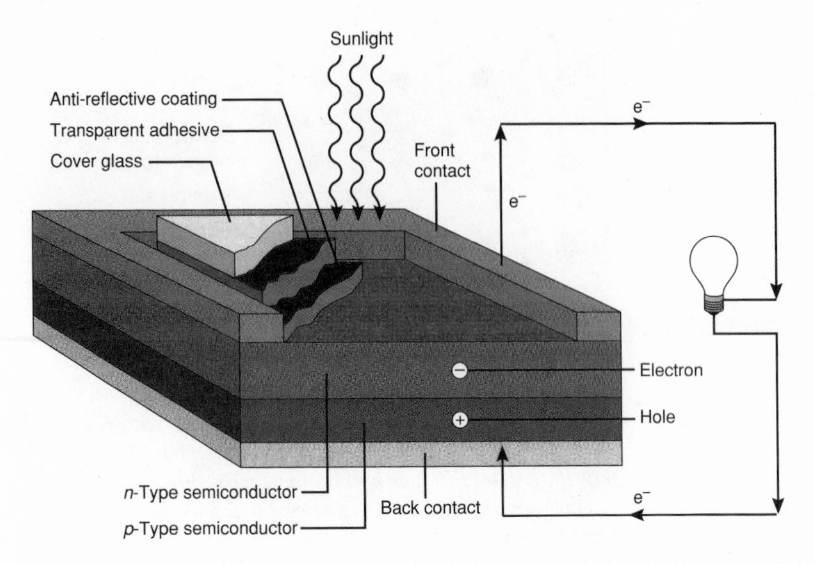

Figure 10.4. The p–n junction is the heart of a photovoltaic cell. It creates the potential difference that enable electrons and positive holes to flow within the crystal. (Reprinted with permission from the American Chemical Society.)

Having described the basic operation of a photovoltaic device we can now better understand why the cost of the end product is so high and what can be done to lower it. Experience with transistors pointed to the fact that silicon was the element of choice for making reliable photovoltaic devices. Although silicon is the second most plentiful element on Earth, it requires a great deal of effort to purify it sufficiently from its natural state. The efficient operation of a photovoltaic cell requires a product which has almost no impurities other than the ones we add intentionally. The silicon used in photovoltaics is therefore very expensive, its cost is about $40 per kilogram. A possible cost-cutting program is to reduce the cost of removing the impurities from the silicon used in a cell.

Silicon found in nature does not consist of a single crystal but of isolated crystals that are oriented every which way. Research on transistors has shown that the most reliable and effective ones, either n-type or p-type are made from single crystal silicon. The best photovoltaic cells are also constructed from single crystal silicon. The electrons and holes have to travel through the bulk of the material before they reach the wires at the surface and they have a tendency to recombine, nullifying their contribution to a current. A single crystal of bulk material offers the best chance for the electrons to reach the surface before they recombine with holes on the way to the surface, and hence produce a larger current.

Large single crystals of silicon have been made using the Czochralski process. A seed of crystalline silicon is slowly withdrawn from a hot melt of the element. The crystal withdrawn is about eight inches in diameter and about three feet along.

To create a photovoltaic cell the block, has to be cut into wafers about a third of a millimeter thick. The slicing is a time-consuming batch operation. Half of the silicon is wasted by the sawing, a serious loss. Nearly 48 percent of the photovoltaic cells now in use are made from crystalline silicon by methods similar to the one described. This method is popular because it produces very efficient devices.

The efficiency of a cell is the ratio of the energy output of the cell to the energy put in by the light source. Presumably, the sun's light energy doesn't cost, so why worry about efficiency? In a real installation, other costs have to be figured into the cost of the energy produced: land, site preparation, cleaning the environment, and the cost of preparing environmental impact statements. Also the size and quality of the structure that holds the system in place must be considered. All of these costs are proportional to the area covered by the system which in turn are inversely proportional to its efficiency.

A considerable effort of the research and development programs is devoted to trading off the efficiency of the photovoltaic device with the efficiency of manufacture. The Czochralski process is circumvented by casting polycrystalline ingots to be cut into the wafers instead of working with a single crystal. Further development of manufacturing processes has attempted to circumvent the cutting process because it is inefficient. Withdrawing sheets of polycrystalline silicon from the melt instead of slowly drawing single crystal rods results in a less efficient photovoltaic device but produces an easily automated, less efficient product whose manufacture cost is less. Photovoltaic cells made from polycrystalline silicon are cheaper to make which compensates for their reduced efficiency.

For some uses, it is possible to save more silicon and still be able to mass produce semiconducting ribbon by spraying crystalline silicon onto some suitable material. The thickness of the silicon is one micron compared to about 300 microns (⅓ mm) for the wafers. Surprisingly, the thinner product is a more efficient absorber of light than the wafer but because there is less material the energy output is smaller. Engineers corrugate the ribbon, which causes internal reflection of the light, and creates an effective, thicker ribbon. Other techniques can simulate a still thicker ribbon. The end result is again a less efficient cell but one that is commercially viable.

Research is also exploring the use of other materials. Gallium arsenide compounds, combinations of gallium and arsenic, for example, make much more efficient cells than doped silicon but its use has been hampered by the fact that the crystals could not be protected against erosion. Silicon is easily protected; when exposed to air, it forms a coating of silicon dioxide. Gallia Inc., a company in Weston, Massachusetts, has found a way to protect gallium arenide by coating it with gallium sulfide.

One of the most surprising and promising developments in this field is the discovery that amorphous silicon, silicon without any crystal structure, can be made into photovoltaic products. Originally conceived in the 1960s by Sanford Ovshinsky, it was rediscovered by R. Chittick of the United Kingdom. Ovshinky's company, Energy Conversion Devices, manufactures amorphous silicon photovoltaic devices and seeks applications for its output.

A glow discharge of a gas, silane (a compound of silicon and hydrogen at 200 to 250°C) can deposit a film of an amorphous form of silicon, called a-Si, on glass, metal, or plastic placed in the discharge. Even though the silicon is amorphous it is a candidate for making photovoltaic devices. The necessary p-type a-Si can be made by adding a gas such as diborane (B_2H_6) to the discharge. An n-type semiconductor results when phosphine (PH_3) is added.

The original amorphous silicon cells, which were only moderately efficient to begin with, had the unfortunate property of becoming less efficient with use. Their efficiencies finally stabilized at about 3 percent. Still they were so cheap to manufacture, that even at this low efficiency they were excellent sources of electricity for hand-held calculators and watches. The Japanese were able to use this inexpensive source of electricity to sell an enormous number of watches and calculators. Today the sale of amorphous silicon is a considerable fraction of the photovoltaic market.

At present the dollar value of the transactions of the photovoltaic industry market is about $1 billion per year. This is impressive for a fledgling industry, but a drop in the bucket of the $800 billion per year for the electrical industry, which is dominated by utility companies. For the photovoltaic industry to succeed in replacing fossil fuels, it must get its costs down.

The utility industry considers photovoltaic concentrator technology a promising substitute for the use of other fuels. In this application the energy of the sun's rays are concentrated a hundred to a thousand times onto a cell, the amplification depending on the light-focusing method and how the sun's movement is followed. The sun's rays are focused on a single efficient cell. By concentrating the rays, much less silicon is used to generate lots of power and thus the costs associated with the size of the plant are reduced. These savings are reduced by the additional costs of the concentrating devices and those required to follow the sun.

This technology is most suited for regions where there is a great deal of direct, cloudless sunshine; diffuse sunlight scattered by clouds cannot be focused. Saudi Arabia has successfully had a 300-kilowatt plant in operation since 1981. Experimental plants in the United States, Japan, and other western countries indicate this is a promising technology.

Utility companies have already been using photovoltaic devices. Pacific Gas and Electric in California have about 700 installations operated by photoelectricity. These are small installations generally out of the mainstream or as backup for regular operating systems.

The home market for photovoltaic devices is not being neglected. New England Electric in Gardner, Massachusetts installed solar photovoltaic panels in homes at a cost of $10,000 per panel. Each panel generates 2.2 kilowatts, sufficient for the power needs of a typical home including heating. The panel occupies about 25 percent of the roof area. The electrical cost and upkeep are negligible and the panel will last indefinitely. It is estimated that the installation cost can be recovered in about ten years, which makes this development almost economically feasible.

To summarize, the worldwide commercial use of energy produced by photovoltaics is only a small percentage of the energy used. Most applications are of the high value kind—in places where photovoltaic systems are economically competitive with traditional ways of providing electrical power. These places include installations remote from electrical grids, in space, and in villages far from an electrical grid. Photovoltaics are also used to replace another expensive form of electricity or in place of batteries. They've been utilized in water pumping, remote communication, emergency lighting, and refrigeration. As the cost of the power from photovoltaic systems decreases and the cost of conventional energy sources increases, more and more niches will be found for photovoltaics. It is also likely this field will continue to develop because every advance will be rewarded economically by additional applications of these devices.

Aside from cost, there is an important drawback to using photovoltaic electricity. Its use depends on sunlight, thus some alternative or storage must be provided when sunlight is not available. Fortunately, the periods of maximum usage of electricity coincide (more or less) with the periods of maximum sunlight so that sun-based electrical generators can have easy access to the energy markets. Should the storage devices be fossil fuel-generated, photovoltaics would be useful in decreasing the environmental problems. Storage devices not fossil fuel dependent are available, the most promising being the manufacture of hydrogen (see Chapter 17).

Some concern has been expressed about the amount of space photovoltaic installations take up. The concentrator technologies have shown that a photovoltaic generating plant need not take up more space than a conventional plant. In urban areas, there are many roof tops which can provide space for an installation.

Some environmental problems have cropped up in manufacturing photovoltaic devices. These are similar to those which have arisen in the manufacture of transistors and have been dealt with using existing technologies. Whether this will continue to be the case as this industry develops and expands remains to be seen.

The prospects for extensive use of this source of electrical energy is difficult to determine partly because the rate at which successful development of the technology will occur is not known and cannot even be estimated. The lowest estimate by the Department of Energy is that by the year 2010 it will supply about 1 percent of the energy use of the United States. This seems much too conservative especially considering there are already laboratory installations where one can estimate a consumer cost of 10 cents per kilowatt hour. There are many places in the United States, such as the east, where fossil fuel-generated electrical power is 12 cents per kilowatt hour.

Prospects seem promising if only because there are vast amounts of energy coming from the sun; there are so many avenues of development available; and the energy costs of fossil fuels are bound to increase. Additionally, this source is renewable and is relatively benign environmentally. No greenhouse danger, no noxious acids, no carcinogenic pollutants come from its use.

On November 15, 1994, in *The New York Times* business pages, the Enron Corporation, a private utility, announced that it proposed to invest $150 million in a plant to generate 100 million watts of electricity using photovoltaic cells. They planned to have this plant ready within two years. The plant has not materialized. Still, such an announcement indicates that using photovoltaics in the largest market for electrical generation is on the minds of the movers and shakers in this business.

In May 1997 the Sacramento Municipal Utility District and the United States Department of Energy signed a contract to deliver 10 million watts of electricity to some hundreds of homes at an initial cost (in 1998) of 16.5 cents per kilowatt hour and at a cost of 8.2 cents per kilowatt hour in 2002, the final year of the contract. The photovoltaic cells will use thin-film silicon, and the panels will be placed on the roofs of the potential customers.

It is difficult not to be enthusiastic about a potential energy source that can drive hand-held calculators, power automobiles (which has been done), or potentially substitute for the energy that large utilities now provide without causing much harm to the environment.

CHAPTER 11

BIOMASS

Biomass is vegetation and organic animal wastes that have been generated in recent times. At this writing about 4 percent of the energy used in the United States is produced by biomass-fueled systems. A little less than half comes from the burning of firewood in stoves used for space heating or for cooking. The remainder comes from corn or sugar used in the production of alcohol (gasohol) as a vehicular fuel, and the burning of municipal solid wastes.

Aside from the United Kingdom which provides less than 1 percent of its energy needs using biomass, the United States is unique among the countries of the world in using such a small fraction. In highly developed countries such as Denmark, Austria, Sweden, or Finland, the percentage is as high as 10 percent. But in developing countries biomass supplies an average of 36 percent of the energy requirements. In India, a more prosperous developing country, the percentage is 56 percent. In Tanzania, it is 97 percent.

The United States Department of Energy hopes that within 20 years biomass will be able to supply between 15 and 20 percent of the energy requirements of this country and perhaps double that in 40 years. The United Nations issued a report in which their technicians maintained the world should be able to have 55 percent of its energy from biomass by the year 2050. One could even be somewhat skeptical of both estimates, and yet be convinced that biomass will ultimately replace a substantial portion of the energy now supplied by fossil fuels.

Despite our wariness of the greenhouse effect that accompanies the burning of fossil fuels, the United States government and the United Nations are sanguine that biomass will not harm the environment. The difference between fossil fuels and biomass is the age. Biomass is organic matter grown in recent times, whereas fossil fuels are the remains of vegetation deposited hundreds of million years ago. When we grow biomass, we remove carbon dioxide from the atmosphere and return it when we use biomass as fuel. It is a balanced operation, neither adding nor subtracting the amount of carbon dioxide in the atmosphere (although one has to be careful about disposing of the toxic chemicals incorporated into the biomass). On the other hand, fossil fuels deluge us with carbon dioxide that was taken from the atmosphere millions of years ago and now return it at a time when it can do us great harm. Incidentally, recently-grown biomass, when burned, delivers fewer toxic substances into the atmosphere than do fossil fuels.

There are various ways of providing the biomass to be used as an energy source. The least desirable way is to use trees that have been growing for a long time.

This is probably what is being used in Tanzania and other developing countries. These practices diminish a resource and cause environmental damage. A more desirable source is to use the residues of ongoing agricultural practices. Bagasse, the residue of the sugar cane crop, has been an important source of energy in ethanol production in Brazil. Other waste materials such as sawdust, scrap wood, or pulp waste from the paper industry can be used. David Hall of University College, London estimates wastes could supply as much as 7.5 percent of the world's energy needs. Other such wastes will emerge as more uses of biomass will be found.

Other promising sources of raw material can be specifically devoted to growing biomass. Various crops can be grown. In Denmark, which is short of forests, they use straw as a source of energy. In Brazil, sugar cane is the crop of choice, which is the source of Bagasse. But we can anticipate that plantations of woody vegetation will ultimately be important sources. These plantations can be on degraded agricultural lands, former forests, or excess crop lands. Revenues from the sale of the biomass could be used to fund the restoration of the very degraded lands.

Burning biomass is one useful way to capture its energy and thus supply either heat or electricity. In 1979 F. Thomas Ledig of the United States Forest Service estimated that burning wood yields 10 to 15 times more energy than the nonsolar energy used to grow it. This estimate includes production, fertilizer, herbicides, harvesting, and transport. Advances since 1979 have made this rate even more favorable. A recent technology calls for converting wood to a gas by heating it in an oxygen free atmosphere. Then the gas is used as an energy source for jet engines. This process recovers 80 percent of the energy originally in the wood.

However, United States hopes to convert biomass to a liquid fuel capable of replacing gasoline. Over 39 percent of the energy used in the United States is for vehicular transportation. Brazil has been successful in eliminating pure gasoline as a means of powering automobiles. In 1975 Brazil found itself almost bankrupt because of the increase in oil prices imposed by the Organization of Petroleum Exporting Countries (OPEC). Brazil decided to quickly substitute alcohol for gasoline in all vehicles. To that end, by subsidies and by tax incentives, it fostered a large increase in sugar cane production and an industry to convert the sugar cane into alcohol. Since 1979 no pure gasoline-powered car or light vehicle can be run in Brazil. Half of the automobiles are powered by pure ethanol and the rest by gasohol, a mixture of 22 percent ethanol and gasoline. The industry is no longer subsidized. Bagasse supplies 100 percent of the heat for the factories and 92 percent of the electricity for the project. Not only has the air quality of the country improved, but the new industry has created 700,000 new jobs.

There are questions as to whether the Brazilian experiment can be duplicated in a country like the United States. Compared to the United States, Brazil has a great deal more sunshine and more unused crop land. Workers there are paid low wages. Moreover, because Brazil has no access to domestic oil, it can produce alcohol at a much higher price and still be competitive with gasoline. Nevertheless, this experiment is a lesson in what can be done to replace a widely used fossil fuel if the government has the will.

Transplanting the Brazilian experiment requires more than substituting one food crop for another. Corn, for example, is not as good a crop as sugar cane. Corn stores its energy in the form of starch rather than sugar. Additional steps are required to convert the starch to sugar before ethanol can be made. The development of biomass as a fuel in the United States envisages plantations of woody plants and an industry that can efficiently convert the woody biomass into a fuel capable of competing economically with gasoline.

To achieve such a goal requires research in a variety of areas. Much of the research will have to be government funded. The economic payoff for this research is too far in the future for the industrial community to be interested in financing it. Besides, the elimination or reduction in the use of fossil fuels is a national objective and is reasonably in the province of the central government.

The creation of a liquid fuel from woody biomass is more complicated than starting from sugar cane. Wood has a cellulose component, requiring a different chemistry for converting it to alcohol. There are a variety of operations for handling the wood, each of which has to be improved over present practice in order to achieve a competitive market price for the end product. Suitable plants have to be grown. They have to be harvested, transported, and stored. Only then can the agricultural product be converted into a fuel suitable for powering automobiles and trucks.

If biomass is to be an important energy source the plants must be efficient in producing an energy-rich end product very quickly. In general broad leaf plants, which have a short growing season and are efficient photosynthetically, are preferred. In addition, these plants should require little or no fertilizer, because producing fertilizer requires considerable expenditure of energy. The plants should be hardy, resisting insect infestation and plant diseases. This would reduce the need to manufacture herbicides and pesticides, a process which also requires considerable amounts of energy. The plants should need little water because water is always a commodity in short supply for farming. To meet these requirements involves a great deal of species selection based on genetic properties. Recent advances in genetic engineering could be very helpful in creating the ideal plants. Considerable progress has already been made in using genetic engineering to change plant characteristics. However, the problems associated with converting biomass to fuel have not been touched. In 1995 and 1996, the United States approved herbicide-resistant soybeans and pesticide-resistant corn, cotton, and potato varieties for farming. Rye that can be grown in aluminum rich soil, soil that ordinarily is wasted for farming, has been created. A great deal has been accomplished but we have a long way to go before the ideal plant is created. Yet there is optimism about the future.

Considering woody plants for energy purposes avoids the ethical questions about the advisability of using natural resources of a country to favor the motor car over the staff of life. Even though the amount of land used for food is declining in the United States, this question is raised every time the possibility of using crop land for energy purposes has been floated. We need not necessarily have to choose between cars and food. As previously noted, woody plantations can be grown on degraded agricultural lands.

The harvested wood will probably not be very large in diameter because we can't wait too long between harvests. The forestry industry has never had to harvest small trees very rapidly. Recently, however, a tractor-mounted device that will reap 1000 trees per hour is being manufactured. This is the minimum rate to produce an economically competitive transportation fuel.

Newly harvested trees contain lots of water. It would be well to eliminate as much water as possible near the site of harvest before transporting the wood. Research on some method for splitting logs in the field to partially dry them before transporting is being investigated. If this process works, standard methods for bundling the residue in place in the field could then be used.

Storage is another area of concern. Organic material disintegrates through decomposition. Ways of slowing decomposition during storage is an old agricultural problem. Perhaps the research done in connection with making transportation fuel from biomass will ultimately help the more general problem of storing food stocks.

A fermentation process is the final step in converting biomass into fuel. Unlike Brazil that manufacturers motor fuel starting with sugar cane, the United States wants to use woody plants. Woody plants that are rich in cellulose must first be converted to sugar before being converted to alcohol. Hydrolysis, the process of converting cellulose to sugars is done using either acids or enzymes. Acid converts the sugar along with undesirable end products that inhibit the fermentation process. The use of enzymes is now the method of choice. Considerable progress has been made in using enzymes in hydrolysis; 90 percent of the cellulose can now be converted to sugar, but the present technology can only convert 5 percent of the sugar into alcohol because the alcohol interferes with the enzymatic action. Research to increase this percentage could play an important role in making alcohol from woody plants an economic competitor to gasoline. Progress has been made. The enzymatic techniques currently available have dropped the wholesale price of ethanol from $3.60 per gallon ten years ago to a little over $1.00 per gallon today. The wholesale price of gasoline is 45 cents a gallon.

In 1989, Chem Systems made an engineering estimate of the ultimate cost of ethanol fuel made from woody plants for the National Renewable Energy Laboratory. This estimate assumed there would be no breakthroughs in technology but anticipated incremental improvements over the course of years. The analysis was complete, including capital costs, with the assumption that the woody residue would be used to generate all of the heat and electricity required for the manufacturing process, and that the surplus electricity would be sold. Their ultimate cost turned out to be about 80 cents per gallon, still not competitive with gasoline at current prices. But if the environmental costs of gasoline are considered, the costs might be comparable. And they will be once the price of fossil fuels increases. The survey warned that many of the assumptions made in the report might not apply to a commercial production environment so it recommended pilot plants be built and used to test the assumptions.

Methanol is another alternative fuel in the alcohol family. It can be manufactured at a rapid rate, and the techniques that currently use coal or natural gas as feed stock, can easily be converted to biomass. The well-known process for making methanol involves heating the feed stock to a high temperature to gasify it. Carbon dioxide and other impurities, if they exist, must be removed leaving only carbon monoxide, hydrogen, and some methane. The hydrogen to carbon dioxide ratio must be proper for a successful run and must be adjusted. At high temperature and pressure, the gas mixture is then passed over a catalyst. The cost of producing methanol is comparable to the (estimated) cost of producing ethanol.

Ethanol is a more desirable transportation fuel than methanol, and the enzymatic conversion is the ultimate method that is being looked at. This process has environmental problems, which have yet to be fully evaluated. The disposal of the spent enzymes, acids, and phenolic wastes have to be considered. For the thermochemical production of methanol the disposal of the flue gases and the partially oxidized residues of the biomass present another problem.

The biomass conversion technology has been well enough established to make a substantial contribution toward replacing gasoline. The environmental problems, still to be faced, do not seem to be formidable. With a large fraction of fossil fuel usage devoted to transportation, is there enough land and water in the world to replace gasoline with a biomass-based transportation fuel? We have indicated there might be enough degraded agricultural land for this purpose. In the book *Power Surge*, by Christopher Flavin and Nicholas Lenssen, the authors estimate that "in industrial countries the land currently withheld from food production could produce just 3 percent of industrial energy use in 1991." While this estimate does not rule out the possibility that one might take biomass transportation fuel development seriously, it does indicate that some inventory of the land and water resources are called for. There are a few other possibilities of replacing gasoline; if biomass won't do it other possibilities must be given a high priority.

The questions raised about using land and water resources for processing biomass have brought some attention to employing bodies of water to provide a biomass root stock. Indeed, two avenues have been tentatively explored. One is to use algae, the primitive microscopic plants that grow in water. The other is to explore the possibilities of using kelp, which grows in sea water.

Algae production has been going on in Japan. The plants must be dewatered before they can be used. Algae can't be processed to produce a useful hydrocarbon until its water is reduced, decreasing its mass by 80 percent. Despite this, a reason for considering algae is that it can grow in brackish and waste water.

Kelp comes in different varieties. Almost all can reproduce their weight in four to six months. The processing of kelp, as well as algae, requires that the harvest be anaerobically digested by bacteria to produce methane, which can be used directly as a transportation fuel. The anaerobic digestion of waste water purifies the water as well. Only one kelp farm, located in Southern California, has been established on an experimental basis. The indications are that the cost of fuel from either kelp or algae processing is comparable to the costs of other ways of using biomass to produce transportation fuel.

If kelp farming were to become a large scale industry, there is a question as to where the nutrients would come from. Kelp growing close to shore can be sustained by run-off from land. For a large scale operation, other nutrients would have to be supplied. It has been demonstrated water can be pumped from the sea's depths to supply the plants, but the environmental impact of this has not yet been properly investigated.

It is unlikely the sea will be other than a local source of renewable energy from biomass. The coasts of industrial countries are dedicated to other uses and the prospects of establishing large kelp farms there are dim, except in island countries such as Japan.

Biomass can be used in other ways to replace nonrenewable fossil fuels without requiring plantations, additional land use, or going into the sea. As mentioned, bagasse, the residue of the cane sugar plant, is used effectively in the Brazilian ethanol industry. There are many other plant residues, which are either burned without taking advantage of their energy content for useful purposes or generally left to rot. As fossil fuels become rare and difficult to obtain, biomass will be used especially in the nonindustrial world. Such a development will not materially affect the prospects for replacing large amounts of fossil fuels with biomass.

Wasted biomass is being experimentally used in Pura Village in the South of India. Virtually all of the energy used in this village is derived from biogas, a combination of methane and carbon dioxide, which is produced by the anaerobic treatment of bovine wastes by bacteria. Methane and carbon dioxide are both serious greenhouse gases, but if the wastes were left alone they would still contribute to the greenhouse effect.

Oily plant crops, which are rich in hydrocarbons, are another possible source of biomass. These hydrocarbons can be easily extracted from the plants. Unfortunately, the oils in these plants are in the form of triglycerides and not directly usable as a fuel. These need further processing before they are suitable for automobile engines. This possible source of biomass energy gets little attention but it constitutes another possibility for those who would try to supplant fossil fuels with biomass material.

The likelihood of biomass replacing fossil fuels depends on success in producing an inexpensive liquid transportation fuel. Brazil's example indicates that when a country faces loss of access to fossil fuels it can mobilize to create a substitute that is both an environmental as well as industrial boon. Whether Brazil's experience is transferable to other less sunny countries with a higher living standard is not certain. One cannot guess what will happen should some other replacement for gasoline become available. One possibility is the use of hydrogen as an energy source. This is discussed in Chapter 17.

There are other uncertainties in promoting biomass as a substitute for fossil fuels. Will it upset the economy because the value of farmland will rise, consequently affecting food prices? Will the economic success of biomass becoming an important source of transportation fuels so drive up the price of food to affect the nutritional condition of the country? Will we make a successful transition from a gasoline-based transportation system to an alcohol-based one considering so much is invested in engine manufacture, oil processing, etc? What is clear is that the technological aspects of using biomass as a substitute for fossil fuels is on its way to being solved. Biomass will play some role in the energy uses of the future; what that role will be is uncertain. The Brazilian experience is a comforting example. Unless a substantial commitment is made by our society to gather enough information to decide what role biomass will play as an alternative energy source, the comfort we may derive from the Brazilian experiment is likely to evaporate. At the moment it seems that the possibility that biomass alone will play a substantial role in replacing fossil fuels is remote unless some research breakthrough changes the picture.

CHAPTER 12

ENERGY FROM WIND AND WATER

WIND

Hurricane or tornado survivors or anyone who has seen pictures of the aftermath of these storms will realize there is enormous energy potential in wind power. The problem is taming this energy so that it is useful without being destructive.

Wind energy is derived from the sun. Because the sun's energy is not absorbed uniformly, some regions of Earth's atmosphere are warmer than others. The warmer air expands and rises; the cooler air contracts and settles. The expanded air is at a lower pressure than the cooler air. The difference in pressure causes a wind to blow. There is a further bonus in wind energy, not attributable to the sun but to whatever caused Earth to rotate in the first place. Earth, in its furious spin, drags some of the air close to it around, causing prevailing winds to blow from west to east.

Humans have made use of this form of energy for millennia. For centuries, ships sailed under wind power until devices were invented which converted wood, coal, or other fossil fuels into the energy necessary to drive a ship.

As early as the 20th century B.C., the Babylonian Emperor Hammurabi planned to use windmills (devices with sails) for capturing the wind's energy and pumping water for irrigation. Notice that despite its end use, we have called the device a windmill. Many centuries later, it was principally used to grind grain between two millstones—hence the name windmill, despite its more general usefulness. Whatever it actually does, any stationary device with sails to capture the wind's energy, is called a windmill. Although it is sometimes called a wind machine or a wind turbine.

Holland made the most spectacular use of wind machines. Descartes, the 17th century French philosopher, said, "God made the world, but the Dutch made Holland." And they did it using the wind.

The Netherlands was virtually uninhabitable as late as the beginning of the 15th century. Most of the country was lower than the surrounding North Sea and when floods came, as they inevitably did, thousands of people died. In 1421, a flood washed 72 villages into the sea. In the mid-15th century, they solved this problem by surrounding sections of land with dikes, which were high earthen walls. They then used windmills to pump out the water to another nearby catch basin. The process was repeated until the pumped out water was close enough to be released into the North Sea. Those pumps worked 24 hours a day, repeating this process for the entire country. Patience was rewarded. By the 17th century the country that is now Holland was reclaimed.

At one time, there were 9000 windmills in that country. These not only had a great utility but added to the picturesque quality of the countryside. In the 19th century, the invention of the steam engine forced the Dutch to shift their allegiance away from wind power. Today there are but 900 or so working windmills remaining. Many of the older ones have been converted to homes (Fig. 12.1).

The Dutch, of course, do not have a monopoly on windmills. Virtually every country in the world has at one time or another used the wind as a source of power. The Dutch introduced windmills into the United States when they settled here. The most popular American model was the light pinwheel type mounted on a high tower and can still be seen on some farms. Millions were introduced into the country to pump water and were later used to generate electricity (Fig. 12.2). When rural electrification was introduced, windmill sales and use declined precipitously.

The wonder is that with a proven source of nonpolluting energy and with the technical skills available in the United States, more

Figure 12.1. One of the many Dutch windmills which have been converted to dwellings. (Photo courtesy of Julie H. Kaplan.)

effort had not been made to replace fossil fuels with wind machines sooner. Even today when, as we shall see, these devices have been improved to the extent that they are as reliable and as economical to operate as conventional power sources, there is no great movement to replace an appreciable amount of coal, oil, and natural gas use with wind power. The country seems to be awash in cheap electricity generated by standard devices using fossil fuels whose capital investment has largely been amortized. Hence there is no great inclination at this time to spend for new facilities.

That is not to say that wind power has been totally neglected. Research on wind machines improvement was stimulated by OPEC's tripling the price of oil in the early 1970s. The first serious reaction to

Figure 12.2. The variety of windmills. (A) The familiar Dutch windmill. (B) American windmill used mostly to pump water on farms but also to generate electricity. (C) Two-bladed windmill used on windfarms. (Reprinted with permission from Saunders College Publishing.)

these price increases occurred in Denmark, not in the United States. There are various reasons for this. The Danish economy was more vulnerable to the upheavals caused by OPEC's action. In addition, there was the United States' prevailing attitude to advances in technology. Gary Soucie, in an article in *Audubon* magazine wrote, "We are a nation of technofreaks who are loath to appreciate old hat ideas and are mostly interested in new technologies; but if someone is at the brink of an abyss, the only progressive step might sometimes be a backward one."

The Danish initiative was rooted in Denmark's tradition of exploring wind power, which dates back to the end of the 19th century. This interest faltered in the 1950s when wind power could no longer compete economically with cheap oil. In the 1970s economic crisis, the Danish government encouraged a shift to wind power again. It not only paid 30 percent of the cost of wind machines for the first nine years but also required utilities to purchase, at fair prices, the electricity generated by these machines.

This resulted in a rapid introduction of wind turbines in Denmark. Currently about 3 percent of the electricity generated there comes from wind machines. It is projected that by the 21st century, 10 percent of the Denmark's electricity will come from the wind. What was of more significance for the United States, as we shall see, was the rapid improvement in wind turbine quality and the development of a manufacturing capability for these machines there. Denmark is one of the world's leading suppliers of wind turbines. Their head start in wind turbine manufacture proved to be a life saver for wind-generated power in the United States.

The United States did not get interested in wind turbines until the 1980s. The confluence of three events led to a "wind power rush" in California. First, a state government report identified three strong windy sites in California as possibly suitable for the establishment of a wind power industry. This coincided with the passage of state and federal tax credits for the development of energy resources not using fossil fuels. At the same time, Congress passed the Public Utility Regulatory Policies Act, which required established utilities to purchase electricity generated by renewable energy sources at approximately the cost of their own generated power.

Wind power farms were quickly established in California. The original wind turbines were technological disasters. Because the original designs tried to ape the propellers of aircraft there was no established industry for supplying these wind machines, thus not one of the machines manufactured was either reliable or economical. In the end the Danish manufacturers came to the rescue. Almost ten thousand wind machines were installed as a transition until a wind turbine manufacturing industry could take root in the United States.

Improvements on these first primitive efforts have resulted in a energy source that clearly has a potential to partially replace fossil fuels. Current wind turbines can respond to a wide range of wind velocities, from 9 to 60 miles per hour, compared to a much narrower range of the original designs. The down times for these machines, originally from 50–60 percent, have been reduced to between 2 percent and 5 percent. This is an important improvement because the episodic delivery of power by a wind machine is one of its principal disadvantages and down times for repairs exacerbate this problem.

Materials have been strengthened. Estimated useful lifetimes of these machines have been extended from five to 20–30 years. The turbines can now be used to drive generators which produce 60 cycle alternating current directly, instead of requiring additional equipment to deliver this most common type of current. Costs of wind-delivered electricity are now comparable to costs of generation from fossil fuel. In 1985 the costs were 16 cents per kilowatt hour. The current cost is about 5 cents per kilowatt hour. This price includes not only the operation and maintenance costs but also the interest on the capital investment, the rent of the land, insurance costs, and all other expenses of running a business. More improvements are on the way. It is expected that costs will be further reduced to 3 cents per kilowatt hour by the year 2000, by using taller towers to capture the winds at higher elevations where they are stronger and blow more regularly. Improvements in the sails and turbines will also contribute to this reduction.

Over 1 percent of California's electrical demand is now supplied by wind turbines. Many states even have a much greater wind potential than California. The Pacific Northwest Laboratories of Richland, Washington estimate that North and South Dakota, using wind power, could supply 80 percent of the United States electricity demand. This is true even if one were to exclude as potential sites 100 percent of the environmentally sensitive and urban lands, 50 percent of the forest lands, and 30 percent of the agricultural lands. These two states are not the only viable sources of an adequate wind supply. The United States has enough wind potential to provide several times the amount of electricity that is currently used. It is unlikely that this form of energy will be limited by the availability of suitable land sites. It is certainly not unreasonable to expect that wind turbines stand ready to provide at least 20–30 percent of the United States electrical needs when the expected fossil fuel crisis finally occurs. Pacific Northwest Laboratories is optimistic about the rest of the world's potential.

Professor W. E. Heronemus of the University of Massachusetts has added to the optimistic prospects for the future of wind generated electricity. Should land use be further restricted, in densely populated countries for example, he has suggested that off shore installations reasonably close to the mainland could supply more than enough energy to satisfy worldwide needs.

The wind is a renewable, nonpolluting, and relatively cheap source of power. In addition wind generators are modular and can be added quickly and inexpensively as demand increases. Fossil fuel generators are concentrated sources and require years to build.

There are some down sides to wind power, none of which seem impossible to overcome. As wind machine use increases there will inevitably be conflicts about land use. Wind power production requires a great deal of space. Because of the turbulence created by the rotating blades, the machines have to be placed between 150–300 meters apart (Fig. 12.3). An efficient wind farm must have at least 100 of these machines. Wind turbines occupy less than 5 percent of the area of the wind farm. The rest of the space can be used for

Figure 12.3. A windfarm in Southern California. (Note transposition of figures.) (Photo courtesy of the Department of Energy.)

agricultural farming, ranching, or any other suitable use while the power is being generated. The real estate experience in Altamont Pass in California indicates that land values may even increase considerably when wind farms begin operating. There are some remote areas suitable for wind turbine use where the value of the power generated could be over 200 times the current market value of the land.

Nevertheless there will be complaints about noise and visual blight. Denmark has addressed these problems by setting standards for appropriate siting to eliminate noise to populated areas. Denmark has also resolved complaints about visual blight by involving the community when examining potential sites of wind farms.

More recently there have been complaints that birds, some of them raptors and other endangered species, have been killed by the rotating blades. In Europe birds are rarely killed by wind machines. In the United States, an average of one bird per month has been killed in California.

The reality is that the potential wind generating capacity in the United States is five times its total energy usage. It is likely that enough remote, sparsely populated windy land (where there are no endangered bird species), lacking competing economic prospects is available. This can make a substantial replacement of fossil fuels by wind a reality.

More serious problems relate to the distribution of the energy once it has been produced. Until enough of this type of energy is produced to warrant creating a grid for its distribution, wind farms must rely on already established grids provided by the existing utilities. Windy parts of the country are far from the centers where most of the energy is to be used. This puts the wind generators at the mercy of the utility companies, who are unfamiliar with the technology and may not cooperate.

Apart from this disinterest (or even hostility) of the utility, there are substantial distribution problems that will exist for wind-created electricity. Small increments in energy can easily be incorporated into the output of existing grids. Should substantial amounts of energy be added to an ongoing distribution system, serious engineering problems would have to be overcome. Some experts think that if the fraction of the added energy is 15 percent or more, the grids would have to be extensively redesigned. Other experts believe this fraction should be as high as 45 percent before redesigning.

The other important problem, not completely unrelated to the first, is that wind-generated power is episodic. This difficulty has been partially overcome by designing modern wind turbines which respond to a wide spectrum of wind velocities, thus decreasing the down times of the generators. To have backup energy available, each wind farm could implement a battery system that stores energy when the turbines are operating. This storage expense must be added to the cost of wind-generated power. It has been suggested that connecting the electricity generated from wind farms widely separated geographically would reduce or eliminate times when no electricity is available. This arrangement would add to the price of the product. There would be cost for constructing these transmission lines connecting different locations. Possibly this problem could be solved by not using the electricity generated by the windmills directly, but by using it to separate water into its constituents, storing the hydrogen obtained, and using that as fuel. The advantages of an energy economy based on hydrogen is described in Chapter 17. It is fair to say, however, that the economics of this alternative have not been worked out.

Should transmission lines be needed either to bring the power to consumers or to connect widely separated generators we will be faced with another problem. Transmission line construction involves a large capital investment. While this would increase the cost of the electricity only by about 1 cent per kilowatt hour, it creates a "chicken and egg" problem. No one will invest in a very large scale energy production facility until the transmission lines are built. And no one will build expensive transmission lines until there are large scale production facilities.

There are few immediate prospects for a quick substantial replacement of fossil fuels by wind-generated energy. The best that can be hoped for is a slow build up of these renewable facilities until the day when it becomes economically viable for the replacement to take place. In the meantime it behooves our government and the governments of other countries to follow Denmark's success by providing economic incentives for producing renewable sources of energy and legally forcing utilities to purchase the outputs.

WATER

Another indirect source of solar energy is hydropower. In ancient times grains were ground by water-driven millstones. In the

Roman Empire there was very extensive use of water power for milling purposes. In subsequent centuries, hydropower continued to be used principally in agriculture. In the mid-19th century a new and important application of this form of energy appeared. The hydraulic turbine which could convert hydropower into electricity via a generator was developed. Today, this is the principal use of hydropower.

Running water is an energy bounty supplied by the sun according to the following scenario. After the sun's energy evaporates water from the lakes, oceans, and rivers, the rains come. When rain water falling on higher elevations run downhill, it can be made to give up some of its energy to power a grindstone, generate electricity, or do other types of work.

Hydroelectricity was first launched in the United States in Wisconsin in 1880. In 1890, the first transmission line for the hydroelectricity was constructed, further stimulating this industry. By 1940 40 percent of the electricity used in the United States was generated by hydropower.

The original hydropower use was by "run-of-the-river" installations. These plants extract the energy directly from the running water. The obvious disadvantage of such an arrangement is that the amount of power available is variable. There can be a glut in the spring, followed by diminished or no power in the dry season. If this power were to be used by a community without auxiliary power sources this difficulty would have to be overcome. Modern installations that generate enough power to satisfy a large community involve the construction of huge dams (Fig. 12.4) where water is stored during periods of a plentiful supply against the time when the flow is small or nonexistent. This solution for providing a uniform supply of electricity creates problems of its own. These dams cannot, of course, be built in very densely populated areas. A substantial population would have to be relocated just to build dams in moderately populated areas. Some of these dams are really huge and the problem of relocation is daunting. In China, the contemplated Three Gorges Project, will involve relocating 1,900,000 people. While most other projects are not of this magnitude, there has been substantial public resistance to large dams in Brazil, India, and Canada. In some countries

Figure 12.4. The Grand Coulee Dam in Washington state, about 80 miles from Spokane. A hydroelectric installation designed to supply electricity to a community is very large. The lake created is also very large and is used for recreation. (Photo courtesy of the Department of Energy.)

legislation restricts the construction of dams in specific localities. Thus, Sweden has four rivers that are protected from having dams built on them. In the United States the Wild and Scenic River Act and the Electric Consumers Protection Act also constrains large dam construction in certain localities.

One possible way to overcome the population relocation problem is to build dams in very sparsely populated areas. The United States does not have such promising sites but other countries, such as Brazil, do. This solution breeds other problems. The influx of tens of thousands of laborers and those looking for work into a sparsely populated region during the seven to ten years needed to build a large dam creates housing problems, and public health concerns. Later, when the dam is finished, it leads to shanty towns.

While there remains an enormous potential for additional hydroelectric power, it is unlikely it will be fully exploited. In the United States there are no longer any exploitable sites for large scale installations. Energy use in the United States has risen appreciably since 1940 but this has stemmed from an increased fossil fuel use. Some retrofitting of existing dams has stimulated hydroelectricity generation but this rise was masked by the increase due to fossil fuel use. Thus the percentage of electricity supplied by hydropower in the United States is now about 9 percent of the total electricity used. The percentage worldwide is about 20 percent.

Hydroelectricity is an attractive method of generating power because it is nonpolluting, renewable, and because there is no fuel cost, quite cheap, considering the large initial capital costs needed to create the facility. There are environmental and societal pros and cons from using hydropower. We have already described some of the cons but there are some pros. It is good for flood control. It provides facilities for recreational use and creates an artificial lake (Fig. 12.4) that can be a source of water for irrigation purposes. One of the largest projects in the United States, the Tennessee Valley Authority, had such benefits rather than the low cost of the power generated as its primary objective.

Apart from the societal matters, there are environmental down sides to constructing large hydroelectric facilities. Their construction generally hurts the region's flora and fauna. It also affects biodiversity. For example, the major portion of the suitable sites in South America are in the rain forests. Exploiting these would require clear cutting the trees. This would increase the carbon dioxide in the atmosphere and affect the animal population and plant life of the forests. There also is the danger of inundating a region should a dam burst. While dams bursting rarely occur, when they do, the results are devastating. In 1975, a dam failure in Henan Province, China caused 200,000 people to drown and many millions to die of famine and disease.

One of the most serious effects in the United States has been the virtual destruction of the natural salmon industry due to the dams in the Columbia River drainage system. Salmon have to migrate upstream in order to spawn, but the turbines and dams on the river are obstacles. Fish ladders built to permit the salmon to go over the dams have not been very successful. Nor has it been successful to transport the salmon around the dams.

While there is no thought of destroying the existing facilities, the prospects for generating more hydroelectricity by constructing large dams is not very promising—although some dams are projected. Some increase is possible by retrofitting existing sites of large dams but this will only dent fossil fuel consumption.

Hydroelectric power could be more promising if one concentrated on the construction of small plants. To realize this potential requires some strong government commitment. China, which has a vast hunger for energy, has made such a commitment. At present, some 37 percent of its total hydropower generation is by small plants. The former Soviet Union may also move in this direction.

The United States has tried to respond to the attractiveness of small plants. The National Energy Act of 1978 and the Utilities Regulatory Policy Act made it possible for anyone generating electricity by hydropower to sell the output to a company at a reasonable price. This move is not likely to have a great impact on the energy market in the United States. However, every little bit helps.

There is one other use for hydroelectricity. Utilities have a problem of being able to supply enough electricity for peak demands. To do this they must have facilities available which are idle most of the time. Utilities which are properly located solve the problem by creating a nearby lake at a higher elevation and having energy generated in off peak hours pump water into the lake. They then meet peak demands by generating electricity with the additional water thus stored. There are about 300 such pumped storage generators in the world, about 70 of these in the United States accounting for about 2 percent of the hydroelectric power produced in the United States.

CHAPTER 13

NONSOLAR ENERGY SOURCES: GEOTHERMAL AND TIDES

GEOTHERMAL SOURCES

We have already discussed one form of energy which is not sun-derived, namely nuclear energy. Geothermal energy and tides are two other forms.

An enormous amount of heat is trapped in the interior of Earth. We are reminded of this by the eruption of volcanoes, or by geysers such as Old Faithful in Wyoming, and by the hot springs that many use to cure ailments. Indeed some estimates indicate that the interior energy is 35 billion times the amount consumed annually on Earth. About 60 percent of this is linked to the planet's formation and 40 percent is heat generated by the radioactive elements still in Earth's mantle and crust.

Only a small fraction of this energy could become usefully available. Economic and technological reasons limit us to exploring only about eight kilometers below the surface. Even so, there is enough energy in this region to deliver a useful source likely to last thousands of years provided it could be tapped.

Lest this estimate seem an exaggeration, we can refer to an article by Stephen J. Gould, a Harvard University professor, in the March, 1996 issue of *Natural History*. He describes the recently discovered realm of living things which subsist on energy not derived from the sun, but derived from the interior of the Earth. In appearance these animals are nothing like those observed on our planet's surface. They subsist on bacteria, which live at elevated temperatures in the absence of oxygen. Professor Gould provides an estimate, originally calculated by Professor Gold of Cornell University, of the mass of the bacteria necessary to sustain the subterranean biota. His estimate makes these bacteria the most massive single group of living things on Earth. Whether this estimate is indeed correct does not detract from the enormous energy supplied by the bowels of the Earth.

Presently, geothermal energy is employed only in places where it has dissipated into the atmosphere through a carrier medium, such as warm water or hot gases. The use of this energy has had a long history. The ancient Romans utilized the hot spring water for bathing and heating their bath houses. Spas that advertised the therapeutic benefits of the minerals contained in the waters coming from the ground were favorite watering resorts in Europe.

Strangely, the first commercial exploitation of geothermal mineral waters was not for its energy but to extract its boric acid. The Etruscans may have used this source of boric acid to prepare the enamels with which they decorated their vases. In the 18th century an industry was established in Monte Cerboli, Italy, to remove boric acid from the mineral waters.

As this industry developed, the energy content of the mineral waters gradually replaced the firewood needed for a variety of processes used in the extraction, and finally for heating the factory.

By early 20th century, geothermal heat became a source of mechanical power and helped replace fossil fuels for heating purposes. Beginning in Italy, the movement to exploit this form of energy quickly spread throughout the world. In 1930 a large scale geothermal space heating system was built in Iceland, an entire island which sits atop an enormous geothermal reservoir. There are 200 volcanoes on this small island. So much geothermal energy is available that 90 percent of the space heating in Reykjavik, its capital, comes from it. Seventy percent of the rest of the country takes advantage of the same resources. Geothermal heat in Iceland is also the source of

Figure 13.1. The Geysers Geothermal Electric Plant in California. It is the largest geothermal installation in the United States. (Courtesy of the Department of Energy.)

steam for industrial purposes, for commercial vegetable farming, and almost all of the boiling of water. Iceland's success caused exploitation of geothermal energy to spread to France, New Zealand, and other countries.

Beginning in 1950, geothermal heat began to be exploited in earnest to generate electricity. The Geyers, California has one of the largest installations in the United States (Fig. 13.1). In New Zealand geothermal heat fuels a paper mill.

The most spectacular way Earth shows its internal energy is through the lava flow from volcanoes. Unfortunately no one has figured a way to exploit this energy. But there are other geothermal reservoirs on Earth where their energy presents itself in a more benign way.

The easiest reservoirs from which energy can be extracted are those which emit pressurized superheated dry steam. Harnessed steam heat can directly drive turbines to drive electric generators. The Geysers, California installation is such a reservoir. It provides a considerable fraction of the geothermal energy in the United States. Such reservoirs are rare. About 20 times more common are wet steam reservoirs which release mixtures of steam and water under pressure. There are still others which release only hot water. These can be made to power an engine to drive a turbine efficiently if the water is hot enough.

A fourth of the emissions of the wet steam reservoir is steam. The rest is water, which contains dissolved in it minerals, salts, and toxic chemicals (such as arsenic and mercury). It is this brine, which complicates the exploitation of geothermal heat. Eliminating hot brine is a problem. Reinjecting into its source in order to extract some more heat from it while disposing of it, poses a risk of contaminating the aquifers. This and other more serious problems would arise were brine merely dumped. Purifying the brine is an expensive operation. There is also the problem of disposing the extracted minerals and poisons. Brine is corrosive. It damages tools, drills, electric cables, and electronic equipment. In many cases brine contains valuable minerals and chemicals. But except for the extraction of boric acid to which we have referred earlier no industry has been developed to do this extraction economically. An economically successful industry of this sort would not only provide for an alternative energy source but would help solve an environmental problem.

These are not the only drawbacks to utilizing wet steam geothermal energy resources. An operating well emits large quantities of hydrogen sulfide, which smells like rotten eggs. This makes the environment around the well unpleasant. So does the noise accompanying the eruption of brine mixed with steam. Finally, many geothermal reservoirs are thought to be finite resources that are exhausted over time. One big problem for their development is that it is almost impossible to estimate how long a particular installation will produce energy. For this reason it is difficult to attract investment capital in wet steam wells.

Other ways of exploiting geothermal energy that might finesse some of these difficulties are available for use in other types of geothermal wells. Reservoirs where only a hot liquid is the source of energy does not necessarily call for introducing this brine into the environment. Rather, the brine is used to vaporize a volatile gas, which then helps power an engine to drive a turbine. The gas is a volatile flammable hydrocarbon. Millions of pounds of hydrocarbon must be circulated to produce an appreciable amount of energy. According to a plant manager, "That's a lot of hydrocarbon fluid to have rolling around." He was implying there was the possibility of either fire or explosion engulfing the plant, which points to the fact that if such a system is used great care must be exercised to prevent a catastrophe.

There are other possible uses of geothermal energy, especially if the liquid emitted is at a temperature below the boiling point of water. Geothermal energy can heat space, produce hot water, supply greenhouses, heat the soil in vegetable farming, or fuel a variety of industrial processes. These uses are possible only if the reservoirs are close to where the energy would be used. It is not possible to transport hot water over a long distance and still have it be a useful source of energy. Iceland is fortunate in that it is a small island rich in volcanoes, a favorite locus for hot water reservoirs, and where the hot water is never used very far from where it came.

Despite geothermal energy's problems, 30–40 countries worldwide use it and are contemplating increasing their use of it. About a fourth of this is for heating and the remainder for power generation. Countries without access to inexpensive fossil fuels stand ready to obtain the most benefits. Those with electricity-generating sources are particularly fortunate because the cost of geothermal-generated power compares favorably with fossil fuel-powered plants. Nevertheless the experts cannot foresee that more than 1 percent of the world's energy needs would be supplied by the described geothermal uses.

The geothermal reservoirs we have discussed are those where the hot liquid or gas escapes spontaneously from the rock fractures in Earth's crust. These sources of hydrothermal energy are only a small, localized fraction of the energy available. Most of the energy in the Earth's crust resides in rocks heated by radioactivity or by the primeval energy source when the Earth was formed. The Handbook of Geothermal Energy published in 1982 estimates there is enough energy in these Hot Dry Rocks (HDR) to supply all of the energy needs of the world for centuries to come. The problem is how to access the HDR which seem to be majestically isolated from the surface of the Earth, where their heat would do the most good.

In 1974 the Los Alamos National Laboratory was issued a patent for their method of mining the enormous amount of heat energy trapped in the HDR. The following description of the process and the status of its development is taken from a series of reports and articles published by this laboratory. The program manager of the project is Dr. D. V. Duchane.

The HDR system requires a shaft to be drilled deep enough to reach rocks hot enough to raise the temperature of a fluid to a high enough temperature to run an engine efficiently (Fig. 13.2). As a practical matter this laboratory envisions the hot water vaporizing a hydrocarbon that would power an engine to drive a turbine that would drive an electric generator. An artificial reservoir is created at this depth by pumping water down under high pressure to open up natural joints in the rock. A second shaft is subsequently drilled nearby to penetrate the reservoir. The reservoir's size can be determined at the surface using microseismic measurements.

During operation, water is pumped down one shaft, across the reservoir, and back to the surface through the second shaft. As it traverses the reservoir, it picks up thermal energy from the hot rock. At the surface, this energy can be extracted and applied for useful purposes. The water which has been cooled in driving the engine is then recirculated in the system.

This process has minimal effects on the environment. The laboratory has been able to have only 7 percent of the water lost.

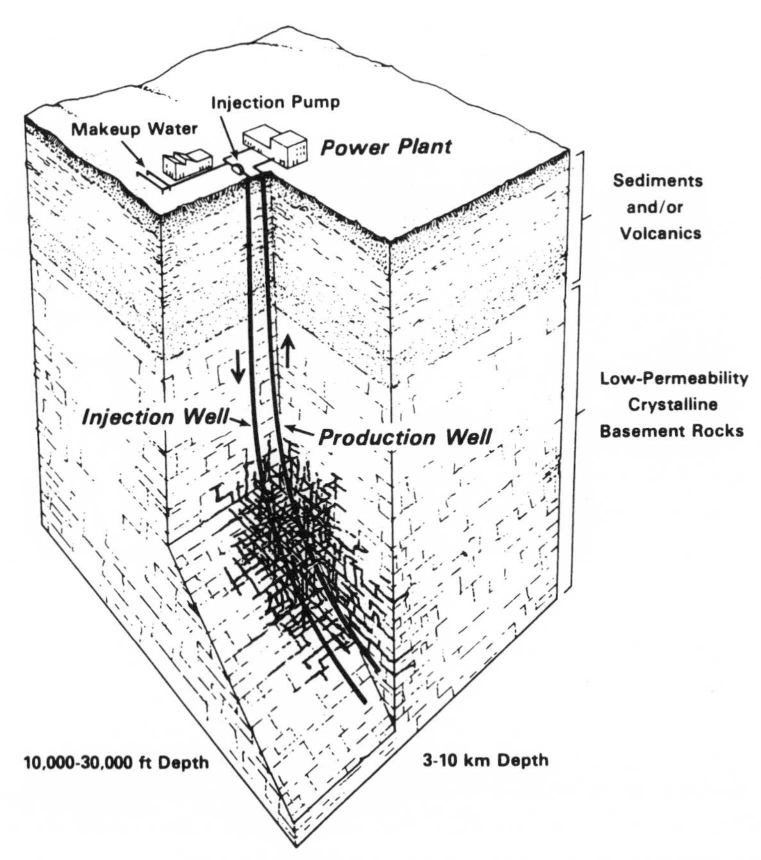

Figure 13.2. The Hot Dry Rock (HDR) System that was at Fenton Hill, New Mexico. (Courtesy of the HDR Project of the Los Alamos National Laboratory.)

Compared to other methods of generating electricity, the installation need occupy only a small amount of space.

Fenton Hill, New Mexico, not far from Los Alamos, has built developmental models that could become electrical generating plants using this system. During this developmental period, various parameters for the optimum operation of the facility have been determined.

The last installation had a reservoir of a volume of 650 million cubic feet. The set-up provided hot water continuously for more than 100 days, with only occasional minor stoppages. Restarting turned out to be easy. Tester and Herzog estimate the ultimate costs are between 5–7 cents per kilowatt hour to generate this electricity.

Other uses are foreseen for the HDR system. It is an ideal method for cleaning up waste water and making sea water potable while generating the electricity. Chemists believe that the hot reservoir could serve as a chemical reactor for producing certain chemical products.

After 20 years of research and development, the Department of Energy invited industrial participation in constructing a facility that would more closely resemble a profit-making source of electrical power. The Department of Energy offered to contribute up to 50 percent of the installation cost of the energy production and marketing facility, up to a maximum of $30 million. More than 40 companies expressed an interest in pursuing this work.

Unfortunately the initial interest expressed by the geothermal power industry did not result in any satisfactory bids. Probably the potential bidders did not trust the cost estimates of Tester and Herzog. Also, because of the Department of Energy's budgetary constraints coupled with the department's threatened elimination by Congress, the incentives offered to industry were probably too modest for such an unproven technology. The future of the project at Los Alamos is in great doubt. Their plant has been closed and the well bores plugged. The Department of Energy wants private industry to pursue HDR research and development. In view of the previous history of such an attempt one cannot be too sanguine about its success.

In the end an industrial project functioning in the real world will have to address some of the issues that a development project cannot. Most important is a good estimate of the capital construction costs. Shafts more than 10,000 feet deep are not cheap. Such a project would also be able to gather the necessary facts to estimate the lifetime of an installation, to see whether the large initial capital costs can be recovered. And finally, the optimal size of the installation and the equipment needed to make it a self-sustaining profitable enterprise, would have to be determined.

Some of Fenton Hill's work was done in collaboration with the Japanese and the European Community. Fortunately, they have excellent sites for such a project and have stated their intention of pursuing it until they have succeeded. The Europeans have projected an industrial facility capable of generating approximately 20 megawatts of power by the year 2002. Of course there remains a giant step between a time table for completing the project and establishing the successful industry.

There is a surprising lack of United States governmental support for technology which, if converted into a successful enterprise, could indeed provide our energy needs indefinitely. Its potential seems as far-reaching as fusion energy. Its development seems well ahead of the fusion project, having had a successful demonstration of a source of hot water that could power a model. Yet is does not seem to attract a fraction of the attention or funding that fusion has attracted. The published progress reports of the project may be too optimistic in describing its potential as an energy source or perhaps it is true we are a nation of technofreaks.

TIDES

Another source of energy independent of the sun is provided by tides. In most regions of the Earth all of the waters, the oceans, the rivers, the bays, the lakes, rise regularly at intervals of 12 hours and 25 minutes, and then recede. The additional energy potential of waters at high tide is clearly a possible source of usable energy, and it is a source that has been exploited.

Since ancient times, the rise and fall of tides has been correctly attributed to the moon's pull on the water. It was not until the end of the 19th century when George Darwin, son of Charles Darwin of evolution fame, provided a definitive explanation of this effect.

It needed to be explained why there are two tides per day since the moon seems to be pulling on the waters of the Earth only directly under it. That would seem to account for only one tide a day. The explanation can only be understood in detail by making an accurate dynamic analysis of the motion of the Earth–Moon system. Because of mutual gravitational attraction between them, the Earth and Moon rotate around their common center of gravity, which is well within Earth's interior. This rotation causes the waters farthest away from the moon to bulge just as someone on a rotating platform tends to be thrown away from the center of rotation (Fig. 13.3). This bulging is the source of the second tide.

The sun also plays a role in the rise and fall of tides. The sun is much more massive than the moon, which would make its influence stronger. On the other hand it is much farther away, which would make it weaker. A careful dynamic analysis yields the result that the sun's effectiveness is 46 percent of that of the moon's. When Earth, moon, and sun are in the same straight line, as they are during full and new moon, the tides are 46 percent higher than they are during the quarters. The strong tides are called spring tides, the weaker ones, neap tides.

Interest in tidal energy was sparked in the 18th century by a book published by Bernard Forest de Belidor. Centuries before this book was published, tidal mills were used in almost all of Europe. The old mills were simple. During the incoming (flood) tide, water would enter a storage pond through a sluice where it would remain until the tide turned around. Then during the outgoing (ebb) tide it would flow back to the sea through a water well (Fig. 13.4).

Small tidal mills can provide some useful amount of renewable energy. To make a dent on fossil fuel consumption, installations have to be large enough to generate electricity. In 1960 at La Rance, in northern Brittany, such an installation was built. This tidal electrical generating plant is still functioning and is likely to last for many more decades.

For a tidal generating facility to be economic, the height of the tide must be at least 16 feet. Many factors affect the height of the tides, the most important are the shape and size of the inlets of the shoreline. Estuaries and gulfs in shallow areas are sometimes favorable locations. Certain shapes and sizes can increase the height of the tides by more than a factor of ten over some less favorable sites. A

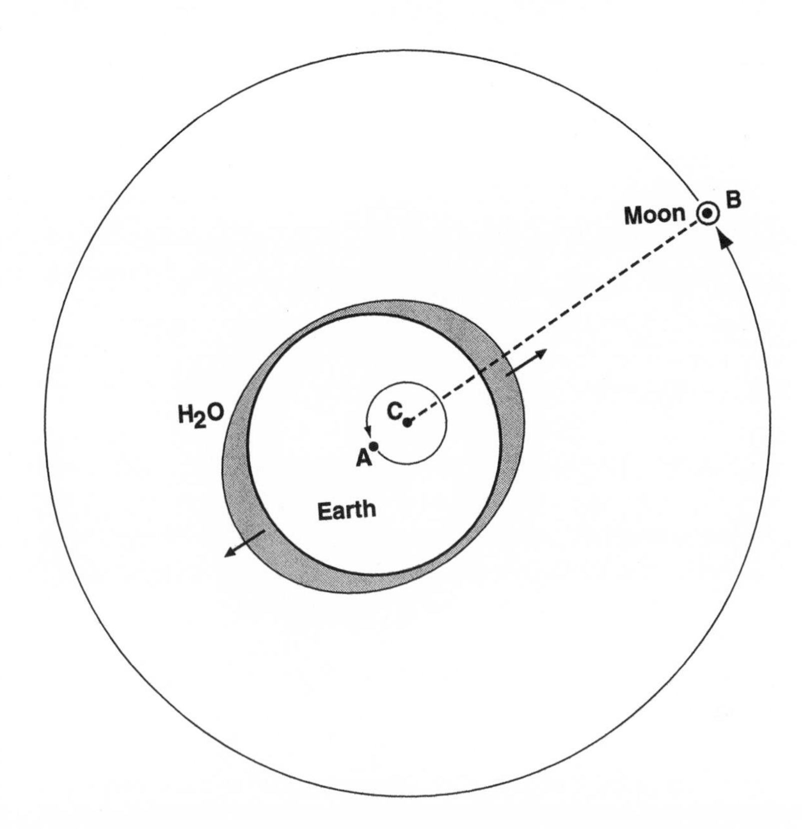

Figure 13.3. The origin of the second tide in 24 hours. Earth and the moon rotate around each along a common axis. The magnitude of the centrifugal force is proportional to the distance from the axis of rotation. It is larger on the side of the Earth farthest from the moon and smaller on the side closest. The gravitational force exerted by the moon is strongest where the centrifugal force is weakest and vice versa. Exact calculation shows that the size of the bulge is the same on each side.

tide can be as high as 40 or 50 feet. Tidal height, and other constraints for the construction of plants, have limited the harnessing of this energy resource.

The difficulties in finding a site for a tidal generating plant can be illustrated by the history of Passamaquoddy Bay and the Bay of Fundy. These are bodies of water shared by the United States and

Figure 13.4. The Tidal Power Station at Kislaya Guba.

Canada. This is the only feasible economic site for a large tidal plant in the United States. In 1935, President Franklin D. Roosevelt allocated $7,000,000 to construct a tidal plant on the shores of Passamaquoddy Bay. Two dams were completed by 1936. The United States Congress failed to continue the appropriations, and the project was completely mothballed because the U.S. Federal Power Commission issued a negative report as to the feasibility of an exclusive United States scheme. Canada still has some interest in developing this resource in some parts of that region but to date nothing beyond a model plant in Nova Scotia has been built.

Electricity generation using tidal power is similar to recovering hydroelectric power. As in old tidal mills, a dam at the mouth of a tidal bay is constructed. A sluice gate permits the water to enter the bay as the tide is rising. A lake forms and is contained by a gate. When the tide recedes the water is permitted to leave the bay through a turbine, which runs a generator. Other more sophisticated schemes for generating energy from tides seem not to be any more efficient.

There are disadvantages to exploiting tidal energy, the principal one is that electricity generation depends on a lunar cycle, which does not coincide with the times of maximum use. The load factor is at most about 35 percent. The best exploitation is to feed the power generated into an established grid provided that it is less than 15–20 percent of the power in the grid. The ultimate cost of the electricity is quite a bit higher than the going cost for power. Depending on the interest rates during the period of construction, costs could cause the price of the power generated to be as high as 18 cents per kilowatt hour.

There are advantages to using tidal energy. The environmental stress is practically zero. Also tidal plants can encourage the development of aquacultural farms, provide recreational sites, and increase tourism.

While tidal energy is not an answer for the United States, there are some locations in the world that are suitable for its practical exploitation. This is especially true for remote or isolated locations where the cost of other kinds of energy is high. Some remote islands with appreciable tidal heads, or isolated parts of large countries like China or the former Soviet Union, are suitable candidates. The engineering technology for exploiting it is well developed. Reliable judgments can be made as to its desirability. Few environmental challenges are posed by this method of energy production. Tidal energy's natural limitations seem to prevent it from making a significant contribution to the replacement of fossil fuels.

CHAPTER 14

OCEAN THERMAL ENERGY CONVERSION (OTEC)

The flux of energy from the sun onto the oceans is so large that if one were to convert as little as 0.01 percent to useful purposes, it would be several times the daily amount of energy presently used by humankind. Converting any of it, however, poses substantial engineering problems. Such a project was not even considered until the early 1970s when OPEC tripled oil prices, forcing the United States to seek alternative energy sources. In recent years the idea of using the solar energy that warms the ocean as a replacement for fossil fuels has languished. William H. Avery and Chih Wu's book entitled, *Renewable Energy From the Ocean* (Oxford University Press) could stimulate new interest in the subject. Much of the information that follows comes from this book.

What makes the recovery of the sun's energy possible is that water is a very efficient absorber of the light energy of the sun. Consequently, the top layer of the ocean, between 15 degrees north latitude and 15 degrees south latitude—i.e. from the Tropic of Cancer to the Tropic of Capricorn—is heated to about 28°C (82°F). The temperature is more or less constant, night and day, in all seasons, and reaches down about 300 feet. Below that level the temperature steadily decreases. At 3000 feet it measures a constant 5°C (41°F) and remains virtually at that temperature no matter how far down we go. This is water that has melted from the polar ice regions, and because cold water is heavier than warm water, it does not mix with the top layer. The cold water layer, like the warm water layer, is also stable, remaining at virtually the same temperature throughout the day and in every season. Many regions of the ocean are much deeper than 3000 feet thus we have a large source of very cold water at our disposal.

As we discussed earlier, engines are devices that convert heat energy into other forms, such as mechanical or electrical energy. We have previously described how an engine works in principle. A brief history of its development might be interesting and pertinent at this point.

A Greek who lived in the first century A.D., Hero of Alexandria, constructed an engine which was little more than a toy (Fig. 14.1). It consisted of a water-containing vessel mounted on a shaft so that it could rotate. As Fig. 14.1 shows, there were nozzles of small diameter protruding from the cover. When the water in the vessel was heated, it caused steam to be ejected through the nozzles, causing the shaft to rotate in a direction opposite to the direction in which the steam was emitted. By this means, the heat energy in the steam was converted into mechanical energy.

There is evidence indicating that the ancient Egyptians and the Chinese constructed engines that could do practical work. But the effective beginning of engine technology came in 1698 in England when Thomas Savery constructed a device to pump water from the mines. Savery's engine was improved in easy steps by Newcomen and Sweaton. In 1763, James Watt invented the greatly improved engine, which marked the beginning of the Industrial Revolution. James Watt is frequently credited with having invented the steam engine. His contribution was substantial but he was not as much an inventor as an improver.

It was not until 1824 when the theoretical analysis of how engines work and what contributes to their efficiency was done. Sadi Carnot, a French engineer, was able to analyze an ideal engine, one without frictional or other dissipative forces, and to determine its maximum efficiency. We have previously summarized Carnot's results; namely that the greater the temperature difference between the hot and cold reservoirs, the greater the efficiency of the engine.

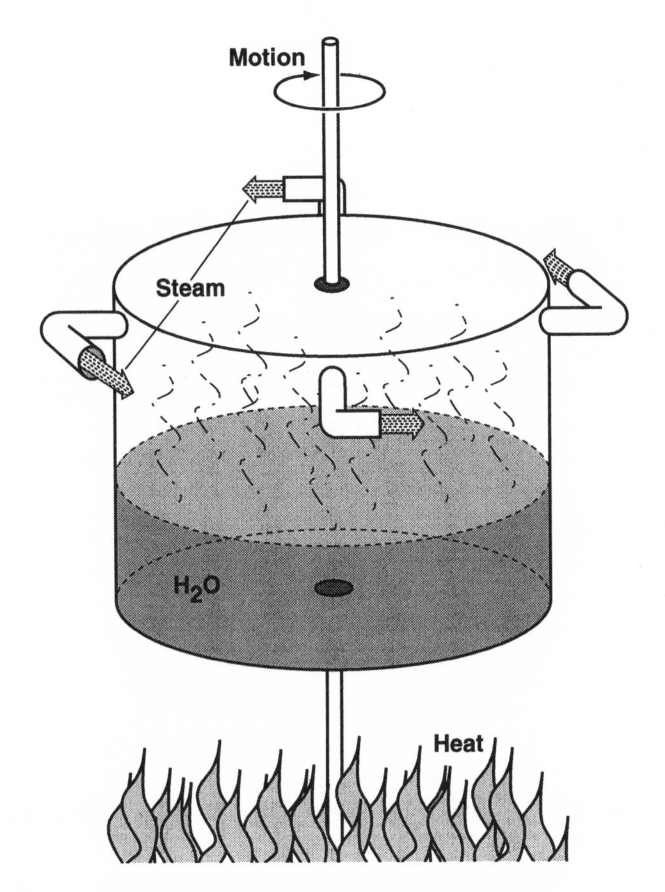

Figure 14.1. Hero's Engine. It looks like a toy but it is a bona fide engine. It converts heat energy into mechanical energy.

The efficiency of Watt's engine was 2.7 percent. This was the proportion of the input of heat energy that was converted into work—a great improvement over the Savery engine, whose efficiency was less than 1 percent. Engine efficiencies are notoriously low but these early models are a far cry from modern engines whose efficiencies can reach as high as 50 percent or more.

Getting back to the matter of deriving energy from the sea the difference in temperature between the top and bottom layers of the ocean suggests the possibility of constructing an engine to drive a turbine and thus generate electricity. The notion of exploiting the sun-warmed ocean as a source of electrical energy was first suggested by Arsène d'Arsonval (inventor of a measuring device for electrical current). D'Arsonval never attempted to realize his concept but one of his students, Georges Claude, was so taken with the idea that he spent a good part of his life trying to demonstrate the practical feasibility of d'Arsonval's vision.

In 1928 Claude successfully constructed a turbine-generator using waste water from a steel plant at 30°C (86°F) and river water at 10°C (50°F). This project's success enabled him to obtain financial support for the next step: to use the cold water of the ocean bottom and the warm water of its surface to generate electricity. His installation off the coast of Cuba was beset by a number of difficulties especially those associated with the half-mile-long pipe required to bring the cold water to the surface. He was never able to generate an amount of power greater than what was required to run his equipment. Nevertheless, he did demonstrate that ocean thermal energy conversion (OTEC) was feasible as a generator of electricity. Claude continued his efforts at least until 1948. His developmental work pointed to the difficulties that could arise and demonstrated, for the most part, how they could be overcome.

After Claude's efforts, no important activity for investigating or exploiting OTEC occurred until 1964 when a detailed scheme was described in a doctoral thesis submitted by James H. Anderson of the Massachusetts Institute of Technology. In those days, there was no economic incentive to undertake a complex, expensive investigation for another source of energy; more than enough seemed to be available at a low enough cost to sustain the United States economy.

The OTEC lead was also not followed up because the efficiency of an engine operating between the two temperatures at the top and the bottom of the ocean was quite small by modern standards. The maximum theoretical efficiency of such an engine turns out to be less than 8 percent. To get a little ahead of the story, when such an engine was finally constructed, its efficiency was 3.5–4 percent. After using some of the energy output to operate the OTEC machinery, the efficiency for generating excess energy was about 2.5–3 percent, which is small compared to a modern automobile engine, while still as good as the Watt engine.

The OPEC (oil) price hike exposed the vulnerability of the United States to the much higher energy costs that could result from crises half a world away. In 1973, Dr. Clarence Zener in his article in *Physics Today*, made an economic and financial analysis of the scheme originally proposed by Anderson. He concluded that OTEC was a viable source of energy, whose cost could potentially compete with the fossil fuels then in use. In the crisis atmosphere of that time, Zener's article was enough to mobilize the United States government. Funding became available for a serious investigation of the problems associated with exploiting the sun-warmed ocean waters to generate electricity.

Stimulated by Zener's article, many United States government agencies began funding projects aimed at determining the feasibility of OTEC as a competitive source of energy. The agencies included the Atomic Energy Commission, the Bureau of Mines, the National Science Foundation, the Marine Administration of the Department of Commerce, and the Department of Energy. Their investigations were thorough and included every possible topic relating to the feasibility of the project including the manufacture of satisfactory heat exchange equipment, the possibility that the equipment would be befouled by the biota dredged up from the ocean depths, and the possible difficulties with engine performance. One of the most important imponderables was agreed upon: the design of a workable cold water pipe as much as 30 feet in diameter that could reach down for more than half a mile and could function continuously in all weather. The platform design, the way in which the energy generated was to be made available to the mainland, and the environmental impact of the entire process were also considered. Most important, an estimate was made of the cost of the energy produced and the conclusion was reached that it could be competitive with energy produced by fossil fuels.

A favorable report was issued in 1975 stating that OTEC was a feasible source of energy. In 1977, essential components of the generator were tested. By this time, the oil crisis had subsided and there was some foot dragging on the part of the government over funding an experimental facility. Frustrated by the administration's reluctance to support development of a complete OTEC system, the Lockheed Corporation, the Dillingham Corporation, and the State of Hawaii collaborated on a project using their own funds to construct such a system. In 1978, only 19 months after the project was begun, this experiment was the first to demonstrate it was possible to run a sea installation to generate an amount of electricity greater than the amount required to run the equipment.

After the success of the Hawaiian project it was obvious that the installation could be scaled up and that a feasible alternative energy source did exist in OTEC. This optimism led to the passage by Congress in 1980 of Public Law 96-310 establishing a national goal to develop solar energy technologies that would be able to supply 1 percent of the energy needs of the United States by 1990 and 20 percent by the year 2000.

Alas, this encouragement came too late. By 1980 the oil crisis had disappeared and the Reagan administration, which took office in 1981, started reducing the government funding. Its assistance dried up completely in 1984.

Interest in OTEC has not disappeared. Many countries are pursuing designs of installations that might some day be constructed. Great Britain is examining plans for an installation in Hawaii. The Netherlands are looking at constructing installations in the Dutch Antilles and Bali. France, the pioneering country, and a Swedish Norwegian consortium are separately considering installations in Jamaica. Governments seem loath to go ahead, and private capital is hard to come by. Japan, however, which is hungry for alternative sources of energy has had a small successful plant operating on the island of Nauru. It is planning a 20-megawatt plant there. The United States is evaluating the possibility of building installations on Puerto Rico and the Virgin Islands.

The planners do not seem to be fazed by the extremely novel installations that would be required. The cold water pipe for a large installation would now have to be about 40 feet in diameter, and the scale of the rest of the plant equally gargantuan. Better estimates are now available for the costs. The installation costs are now estimated between \$2000–\$3000 per kilowatt, about the same as a nuclear plant, and the ultimate cost of the electricity at about 7 cents per kilowatt hour.

Surprisingly, the energy produced with such a low efficiency generator can compete well with more conventional sources of greater efficiency. There are a variety of reasons why this is so. The primary one is that the fuel is free. At the moment there is no land cost, although if the installations fall within the territorial waters of a country, there will be taxes. For installations outside of any country's boundaries, diplomatic negotiation may be necessary to settle such matters as to who can build where, but at the moment there is no cost (such as rent). The low pressures and temperatures in the installation make for equipment that can operate reliably with low maintenance costs. Moreover, plants have been designed to operate without personnel for long periods of time. OTEC enjoys other advantages over conventional sources of energy. For example, no arguments are likely over the siting of an unsightly plant. Apart from the construction of the cold water pipe, few innovations need to be explored in constructing the system. Even the procedures for constructing ocean platforms able to withstand waves and storms are well known as a result of the decades of experience gained from building oil platforms.

There are two modes of generating the energy in the OTEC installations. In the 1st mode, the open cycle, sea water is used as the working fluid of the engine. The warm sea water is introduced, a vacuum is created so that the sea water boils and in its expansion does some work and cools. The steam is then condensed by cold water drawn up from the ocean bottom. This condensed vapor is then discharged into the ocean and the cycle is repeated. If the water vapor is condensed by using cold water to chill it in a heat exchanger,

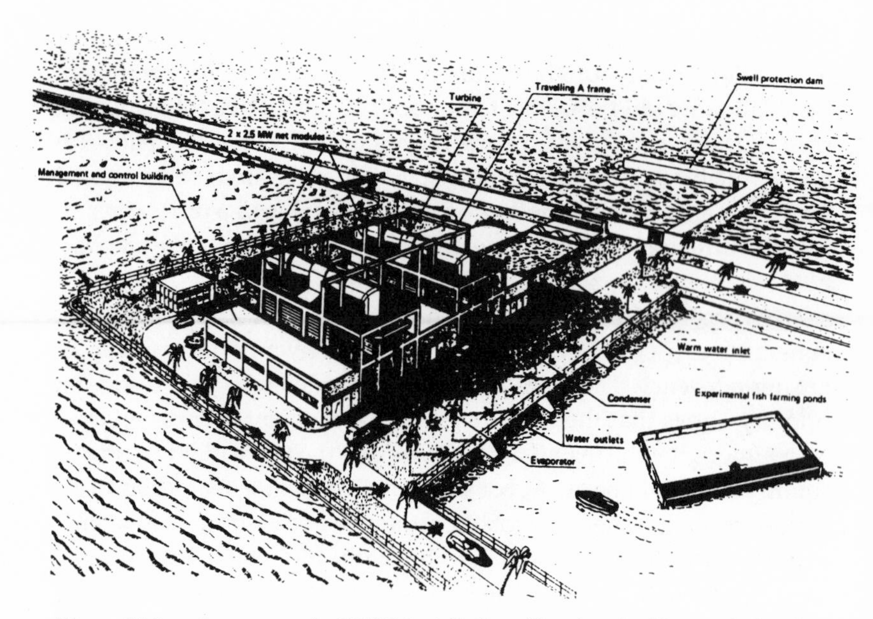

Figure 14.2. An open cycle OTEC installation. (Reprinted with permission from the American Society of Civil Engineers.)

the facility becomes a source of drinking water, an advantage in some potential installations. An open cycle plant is drawn in Fig. 14.2. There is not enough detail in the drawing to show the important characteristics of an OTEC facility, but it is presented to show that there is a working model of the plant and that it is complex.

In the second mode, the closed cycle, the working substance is a volatile liquid; ammonia is the one most frequently used but others are being investigated. This is first vaporized and subsequently cooled in a heat exchanger. The gas does the work upon expansion as is usual in an engine. The working fluid, the ammonia, is reused. At present the open cycle system is more highly developed but the closed cycle mode will soon be able to compete.

Hawaii's experience has shown that there are additional advantages in the operation of OTEC. Cold water can be used in warm climates for air conditioning. Being drawn from the bottom of the ocean it also brings with it a large supply of nutrients that can be used profitably for the culture of saleable aquatic forms of life. And in many communities the possibility of creating a convenient source of clean water would be of great advantage.

A great deal of thought has been given to the question of how the energy generated could be delivered to where it is to be used. The most obvious solution would be to transport the electricity directly by cable to some mainland point and distribute it from there. This is especially feasible in some islands where there are deep sea trenches close to shore. If OTEC is to be used on such islands, the installation could even be on land close to the shore. If the generator is on land or in the sea near the shore the transmission would not pose any problem. In short, tropical islands might be a useful place to test the feasibility of ultimate large scale use of OTEC.

Other suggestions for using the energy generated by OTEC abound. Electricity can be used to convert alumina to aluminum, currently a large consumer of electricity. Some have suggested shipping nitrogen from shore and converting hydrogen (which could be obtained by dissociating water with the electricity generated) into ammonia. This then could be shipped to shore where it would find a variety of uses, especially in the manufacture of fertilizers. But the most intriguing use of OTEC energy is to make hydrogen and to use the hydrogen directly as a source of energy (see Chapter 17).

The least expensive use at the moment for the hydrogen generated by the electrolysis of water using the electricity generated by OTEC is to turn it into methanol. This hydrocarbon can be used directly in automobile engines. Because there is thus an immediate market for the energy generated by OTEC, it has been possible to estimate the ultimate cost of the methanol and compare it to the cost of gasoline and natural gas. Avery and Wu do this analysis in their book. They concluded that ultimately the price of the methanol would be comparable to the current prices of gasoline but when it would become cheaper depends on the interest rate on the money borrowed to construct the plants, how much methanol produced, and the amortization rate of the start-up costs. Unfortunately, using methanol contributes to the greenhouse effect. But this estimate was made to show that OTEC offers considerable promise of being able to supply renewable energy that is competitive with energy from fossil fuels.

There are, according to the United Nations, 53 islands in the tropical areas that might be candidates to benefit from OTEC installations. If one were to include the benefits of a clean water supply and marketing the shellfish and other nutrients dredged up by the cold water pipe, installations in these islands would be economic bonanzas. There are also sites with deep water distant to volcanic islands where transmission costs of the electricity generated would contribute to a high cost per kilowatt hour. These installations might be used economically as incremental power sources in periods of large demand.

The environmental problems that OTEC may present have not been fully explored. Dredging up the cold water from the deep will release some carbon dioxide which is dissolved in the water. However, the amount of this gas released would be only 125th of that released by burning coal to produce the same amount of energy. The cold water may affect the climate and the fauna of the region where the OTEC facility is located. It may also affect the marine life. Finally, just as with the growing of kelp, no one knows the ultimate effect of continually denuding the depths of the ocean of nutrients.

It is hard to know the ultimate importance of OTEC. Some authorities believe that Avery and Wu have demonstrated OTEC could be an important renewable energy source and that its development should proceed rapidly. The World Watch Institute thinks OTEC presents unforeseeable problems. Others believe OTEC has limited usefulness for those parts of the world where the costs of alternatives are high. The final verdict will be based on economic considerations after more experience is gained from observing actual installations and after some of the environmental problems have been further investigated. The inviting prospect of directly capturing the cornucopia of solar energy to use in our daily lives—without environmental penalties—should keep interest in OTEC projects alive.

CHAPTER 15

BATTERIES, FUEL CELLS, AND FLYWHEELS

BATTERIES

Cars and trucks are responsible for using almost 30 percent of the fossil fuel energy consumed in the United States. Almost all of this energy comes from petroleum products. When gasoline and diesel oil is burned, it emits, as a byproduct, ozone and nitrogen oxides. Nitrogen oxides are largely responsible for the smog in large cities, and ozone is an irritant of the lungs, a serious irritant for those who have breathing problems.

There is a twofold purpose in seeking substitutes for fossil fuels in motor vehicles. One is to find a renewable source of energy in place of petroleum products. The other is to mitigate the environmental damage caused by the burning of gasoline and diesel oil. The serious harmful environmental impact caused by burning fossil fuels has not been adequately dealt with by society. The smog problem and the pulmonary irritation due to ozone have finally aroused the public and have led some state legislatures to propose serious measures for partially replacing automobile gasoline engines with electrical batteries.

Batteries are devices that convert chemical energy directly into electrical energy. While they solve the emission problem, they do not necessarily replace fossil fuels. Energy cannot be created out of whole cloth (unless the cloth is burned); the energy available in a battery must be supplied from some other source. At the moment that source is likely to be fossil fuels, although using a renewable source for this purpose is not out of the question.

Until the early 19th century the only known electrical phenomenon was static electricity (electricity without any moving charges), and thus incapable of doing work. Electricity with moving charges (electric currents) was first produced by Alessandro Volta in 1800. He did this by inventing a chemical electrical cell. A number of these cells combined, as is usually done, is called a battery, in analogy with a combination of cannons, also called a battery. Electrical energy can only be utilized when it appears in the form of currents. Volta's contribution to science and technology was a large one.

Volta had no idea he was on to a seminal discovery. He was trying to refute an assertion by Galvani, a fellow scientist and a compatriot. Galvani had inserted two dissimilar metals into the muscle of a frog's leg and found that it twitched. He thought the frog's reaction was due to a "life force" in its leg. Volta thought that if a frog's leg could be made to twitch, he could introduce two dissimilar metals together in an inorganic setting, find the same result, and bring the "life force" explanation into question.

What Volta did was to separate pieces of zinc and silver with a piece of cloth or paper, which had been previously soaked in a salt solution or an acid. He then measured the voltage, or electrical pressure between these two metals, a measurement that was accessible to those acquainted with static electricity. He found that such a voltage could cause the frog's leg to twitch, thus demonstrating that the twitching did not require the existence of a "life force." Volta also connected a number of his cells together and found that by doing so he could increase the voltage of the combination over that of a single cell. This was the first battery.

The direct conversion of chemical energy into electrical energy (which is what a battery does) is possible because atoms and molecules are electrical in nature. They consist of negatively-charged elementary particles, electrons, which are very mobile, and positively-charged ones, protons which are less so. To illustrate how a chemical battery works we will describe one that is not used in practice, but is simple to understand.

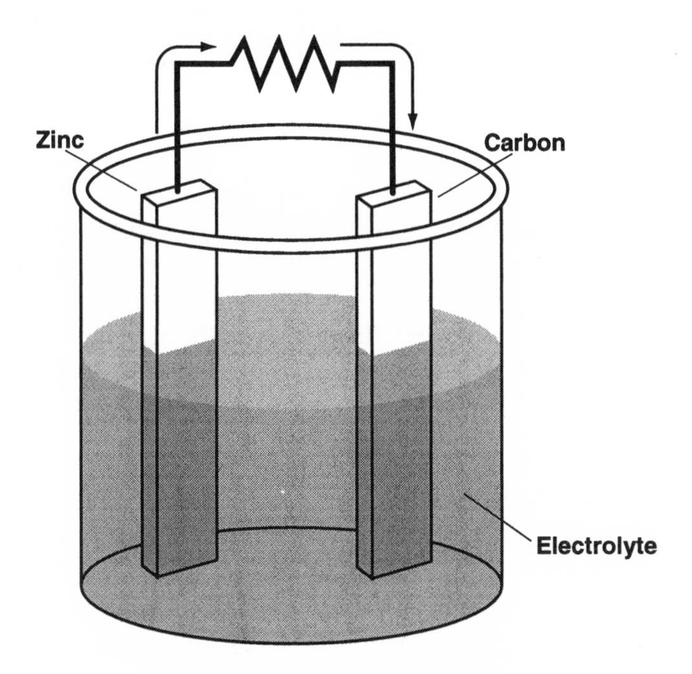

Figure 15.1. A zinc carbon battery.

Rods of carbon and zinc, called electrodes, can be inserted into a bath of some acid, called an electrolyte (Fig. 15.1). The acid attacks the zinc, which then enters into solution, not as a zinc atom but as the atom minus two electrons. This leaves the zinc rod negatively charged. The zinc atoms minus the electrons are called ions. They are positively charged. As the acid becomes positively charged because of the zinc ions, electrons are pulled off of the carbon rod because of the attraction of electrons to positive charges. The carbon rod becomes positively charged. If a wire is placed between the carbon and the zinc electrodes, connecting them, the mobile negative charges on the zinc electrode will migrate to the positively-charged carbon rod, creating a current of electricity. This current can light a lamp, run a motor, and the like. As the current continues to flow the zinc is consumed and the battery dies. The voltage of the battery depends only on the chemical constituents of the cell and not on its size or anything else.

Batteries are used in various ways but they are most useful whenever a portable supply of electricity is needed, such as in automobiles. The lead acid battery with electrodes of lead and lead oxide and sulfuric acid as an electrolyte is the usual configuration for this application. Neither electrode is consumed as the lead acid battery is used. When, as a result of the chemical reactions, both electrodes become lead sulphate, the battery dies. The useful feature of the lead acid battery is that by passing an electric current through it in a direction opposite to the direction it takes on discharging, the electrodes can be restored to their original condition and the battery can be reused. This recharging and discharging can be done thousands of times without the battery wearing out.

Many other types of battery are possible besides the two we have described. Because of the intense interest in using batteries to power automobiles efficiently, numerous different configurations of electrodes and electrolytes are being investigated. It is a lively area of research but has yet to produce breakthrough results.

Batteries could have a powerful effect in improving the environment if they could replace the gasoline used in automobiles. Electrically-powered vehicles have been used successfully in golf carts, warehouses, and factories. They have advantages over gasoline engines in many respects. They are nonpolluting in operation. (Their manufacture may pollute the environment. However, it is easier to control the pollution produced by a factory than the pollution coming from the tail pipe of an automobile.) They are 70–80 percent efficient including the electric motor for driving the vehicle, compared to the 20–30 percent for a gasoline engine. An electric motor is simpler, lighter, and easier to maintain than a gasoline engine.

There is another advantage in using batteries to power automobiles. An average of 20 percent of the energy supplied by a gasoline engine goes to stop the vehicle by using the frictional forces applied to the brake linings. This frictional energy is dissipated as heat and does not contribute anything to propelling the car. When an electrical system is used as a power source, it is possible to flip a switch and reverse the current's direction and slow or stop the vehicle. This works as well as a conventional brake but with this difference: the energy of the moving car is converted to a current which adds to the charge of the battery and is not wasted but can be used later to propel the vehicle. This contributes greatly to the overall efficiency of an automobile powered by batteries.

At the moment, however, the disadvantages of a battery outweigh its advantages. It is heavy for the amount of energy it is expected to deliver. A battery that would deliver the same amount of energy as a filled gasoline tank in a typical automobile would weigh several tons. Moreover, the energy can only be slowly delivered. It is like a water bottle with a small opening. The contents are accessible but can only be poured and refilled very slowly. This translates into an electrically-driven car that can't accelerate quickly and takes hours to recharge. Finally, at present, a typical battery-driven car can only travel fewer than 100 miles before the battery needs recharging. This is still not the end of the story, however. Recent experiments indicate that it might be possible to build batteries that could drive automobiles greater distances before they need recharging. It is expected that these new batteries also could be recharged much more quickly than the lead acid battery.

Despite these defects, California, New York, and Massachusetts passed legislation which required, starting in 1998, 2 percent of the automobiles sold by manufacturers be battery driven. These automobiles can be used for errands within the city. These states expected this legislation would stimulate automobile companies to conduct intensive research to improve the energy content of the battery, increase the time between recharges, and extend the time between recharges. An electric car would then become a more desirable vehicle, thus increasing the demand for these cars so that the 2 percent quota could be sold.

The major automobile companies have maintained that the market at this stage of technical development of the battery-driven car could not sustain a 2 percent volume of sales. California has been convinced by this argument and has rescinded this demand. Instead the automobile companies have promised to continue developing not only these vehicles but other "Zero Emission Vehicles (ZEV)" and to sell them at a rate dictated by the market. To show good faith, General Motors began in 1996, to sell a battery-driven car with a sale price in the middle $30,000 range. This corporate decision is expected to result in a considerable financial loss for this car. Yet other companies, Toyota and Chrysler, have already announced plans to market similar cars. A small company in California, Zebra Motors, has a model of an electric car which it claims can be sold for $19,000 by a dealer at a profit. Another indication that battery-driven cars might catch on is in Sacramento, California. This city has purchased battery-driven buses to replace some of their gasoline-powered vehicles. They were able to overcome some of the difficulties in using batteries because they could install an overnight recharging facility on their premises.

California has issued standards for a "Zero Emission Vehicle" with the requirement that at least 2 percent of manufacturer's sales by the year 2005 must satisfy these standards. What they had in mind was that these should be battery-operated cars. Since then there has been a flurry of activity on the part of manufacturers to develop products that could reduce drastically their emissions. Honda has developed a gasoline engine coupled with a catalytic converter that meets the specifications of a "Zero Emission Vehicle." Toyota is about to market a hybrid car with a small gasoline engine (therefore less polluting) which will charge batteries used to drive the automobile. Arthur D. Little, in collaboration with Chrysler, has succeeded in developing a fuel cell that can produce the hydrogen necessary to generate the electricity by using gasoline as a fuel. This would not only be less polluting that standard gasoline-driven cars but would have the advantage of being able to use the network of gasoline stations already in existence.

The governor of New York publicly indicated that he is in favor of following California's lead. It is unlikely that Massachusetts will stick to its resolve in view of these developments.

The January–February 1998 issue of the magazine *The Sciences* has a description of a trip taken by an experimental battery-driven car that might give some indication of what the future of such vehicles might be. The automobile developed by the Selectria Corporaion of Wilmington, Massachusetts was powered by a nickel metal hydride battery. It was driven 211 miles from Boston to New York City at an average speed of 55 miles per hour on major highways without requiring a stop to recharge the battery. It was equipped with accessories such as air bags and a compact disc player and had a passenger who described himself as a man of hefty proportions. It consumed less electricity than it takes to cook a Thanksgiving dinner, twenty-eight kilowatt-hours.

The battery-driven automobile makes a contribution in reducing air pollution. Such a car has a limited specialized use. Even if it could be marketed successfully it is unlikely that it will completely replace the gasoline-driven vehicles. It is unlikely to make even a small splash unless it catches on sufficiently that it can be economically viable for service stations to recharge batteries and do repairs. If research in battery development produces a battery that could overcome economically some (or all) of a battery-driven car's defects, its future would have to be reevaluated. Interest at the moment is in developing a hybrid automobile that uses a gasoline engine to either recharge the battery or kick in when the battery cannot perform adequately. This would not get rid of our dependence on fossil fuels but would serve as a conservation measure because an engine used occasionally or used to recharge a battery is smaller than one used to power an automobile. Hybrid vehicles of this sort have been developed which can travel 80 miles on a gallon of gas consumed. Plans by many companies for marketing such vehicles soon are proceeding rapidly.

The automobile companies are presently jointly engaged in a research and development project called, "Partnership for a New Generation Vehicle" (PNGV). Their principal objective is to design and market an affordable fossil fuel vehicle which is three times as efficient as the best of the current lot. The project is being supported by every government agency that could possibly have an interest in it. The automobile companies expect to be able to market the results of their work by 2004. This is a laudable attempt as are other conservation measures (as we will see in the next chapter), but it does not address the serious environmental hazards posed by burning fossil fuels.

Since reducing but not eliminating the harmful effects of fossil fuels seems to be an accepted public policy, it is surprising that the attempt to replace gasoline engines with ones driven by natural gas has not succeeded better than it has. As has been noted natural gas is the least polluting of the fossil fuels by a wide margin. It is a cheaper and more efficient power source than gasoline. It requires a minimum alternation of a gasoline engine to use it. It was given a push toward acceptability when governments converted their fleets to natural gas-driven vehicles. But greater acceptability has occurred minimally recently.

Some large companies, such as Consolidated Edison Company of New York, have converted part of their fleets of trucks to natural gas use. These companies buy large quantities of natural gas and store it in their facilities. Therefore, they do not have to rely on the availability of service stations.

The Ford Motor Company has produced Crown Victoria models that can use natural gas as a fuel. It is offering financial incentives to buyers. The New York State Research Institute is offering financial incentives for converting existing automobiles to natural gas-driven ones. And a taxi garage in New York City has just purchased 100 of the Ford vehicles to replace their entire fleet of taxis to natural gas use. Some automobile companies, notably Honda, have announced their intention of marketing cars propelled by natural gas soon.

FUEL CELLS

An alternative power source, which can not only improve the environment but also eliminate fossil fuels as a power source, is the fuel cell. Like a battery, a fuel cell converts chemical energy directly into electrical energy with no harmful byproducts. The fuel cell was invented in 1839 by Sir William Grove, although there are indications that, in the previous century, Sir Humphrey Davy understood the principle by which the fuel cell operates. It was little more than a curiosity until a century later when it was found to be an extremely

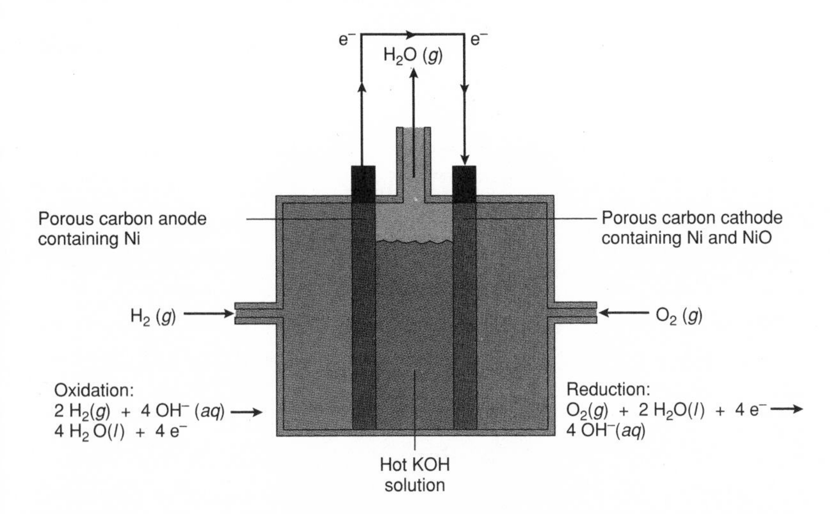

Figure 15.2. A simple fuel cell. (Reprinted with permission from the American Chemical Society.)

useful device for powering the electrical equipment of space vehicles and providing the astronauts with water at the same time.

There are many different designs for fuel cells that are still being researched. However, we will illustrate the principle using a design similar to one originally used by NASA (Fig. 15.2). The idea behind this particular fuel cell is to channel into electrical energy the chemical energy that is released when hydrogen and oxygen combine to form water. Hydrogen is strongly and rapidly drawn to oxygen. When these two are in the presence of platinum or a spark, they release energy explosively. The fuel cell accomplishes this same reaction slowly, without an explosion, by transferring electrons from hydrogen to the oxygen via an external current that can be made to do work.

The cell consists of three compartments formed by two porous carbon electrodes, one containing nickel and the other nickel oxide. The nickel and nickel oxide act as catalysts to facilitate the basic chemical reactions of the fuel cell. The compartment between the two electrodes is filled with an electrolyte, a potassium hydroxide solution. The electrolyte is kept hot to improve its conductivity. The two electrodes are connected by a wire.

In the operation of this fuel cell, hydrogen is fed in at one electrode and oxygen (really air) at the other. When the hydrogen hits the electrode, electrons are stripped off and these can then carry a current through the wire to the other electrode. Electrons make this journey because they are attracted to the oxygen just as they would be if the hydrogen and oxygen were mixed in a test tube. Chemical reactions go on at each electrode. The electrons at the first electrode are ultimately replaced by the migration of a negative ion in the electrolyte. The net result of the chemical reactions in the fuel cell is to have the hydrogen combine with the oxygen to form water just as they would if they were in contact with one another. The amount of energy created in the current is almost the same as if the two chemicals were combined explosively, but here the energy is available at a rate that enables it to do useful work. The voltage of the fuel cell we have described is a little more than one volt, but fuel cells, which are very small, can be stacked to increase both the voltage and the currents created.

In addition to being used in space vehicles, fuel cells have been used to generate electricity for large communities Fig. 15.3 (for parts of Tokyo, for example) and to run buses. They have not been developed sufficiently to be used in automobiles. The fuel cell runs most efficiently with hydrogen as a fuel. Hydrogen is not readily available as a gas because it is expensive and because storing it is a problem. Equipment must be added to the fuel cell to extract hydrogen from another gas, usually methane. For this reason the total equipment is too heavy for use in automobiles but is satisfactory for buses. However, researchers trying to improve the fuel cell are testing, among other things, different elements to make it lighter. There is reason to hope that such research will succeed. Fuel cells may be given an opportunity to test their usefulness in automobiles. As we have noted, the Arthur D. Little laboratories have invented a system of supplying the hydrogen necessary to operate the fuel cell using gasoline. This innovation may be suitable for passenger automobiles and may help retain the gasoline distribution system we now have.

Figure 15.3. Fuel cells can not only be used to drive motor vehicles but can also be combined to provide enough power to light a village or a city. (Reprinted with permission from Saunders College Publishing.)

Batteries and fuel cells are not renewable sources of energy, although they tend to use materials which are in great supply. If methane or gasoline is used to produce hydrogen for fuel cells, they cross-the-line into the nonrenewable fossil fuel category. The possibility exists, however, that hydrogen from renewable sources might become available at some future date. If fuel cells could be manufactured for nonpolluting, renewable sources, they would be ideal. Fuel cells have an added advantage. They can power a vehicle using an electric motor that is quieter in operation than the engines that normally power a bus or an auto. Thus, they would help mitigate the noise pollution associated with vehicular transport.

FLYWHEELS

Finally we turn to energy that can be provided from a seemingly unlikely source—flywheels. These are rapidly turning wheels that store a great deal of energy. They have been successfully used in the past on an experimental basis for powering cars. The energy of the flywheels must be supplied by some external source, probably electricity, which could be generated either by a renewable source or a fossil fuel.

This old idea is being considered because recent improvements that increase the tensile strengths of solid materials may extend the usefulness of flywheels, or indeed be the primary source of power to drive motor vehicles. The larger the wheel and the more rapidly it turns, the greater the amount of energy stored. The periphery of a large, rapidly turning wheel imposes great internal stresses on the wheel. It is the improvements in the kinds of materials recently developed for making these devices that are responsible for renewed interest in this idea.

A flywheel's energy would be dissipated if left alone because of the friction of the air around the wheel and because of the friction in the bearings. Modern technology has been able to address these problems successfully enough that flywheels could be considered as effective energy storage devices able to drive a car.

New developments in carbon compounds composites have made it possible to fabricate a 25-pound wheel that can be made to rotate at a speed of 30,000 revolutions per minute without flying apart. The wheel can be housed in an evacuated chamber to reduce or practically eliminate air friction. (This chamber can also serve as a safety device should the flywheel come apart.) The only remaining frictional source is in the bearings, and this can be made very small indeed.

A flywheel can power a car or a truck using the following strategy. Magnetic material is embedded in the wheel that is made to rotate inside a coil of wire. A rotating magnet inside a coil is a standard way of generating a current in the coil, which is then used to drive a motor that can turn the car's wheels. Sending a surge of current through the wire imparts energy to the wheel and makes it rotate. Plugging into a source of electrical energy to make the flywheel rotate is the equivalent of filling the gas tank of a conventional automobile.

Two ways of using such a device are being tried. One is to use the flywheel in a battery-driven car to supplement the battery. The flywheel can either keep recharging the battery to prolong its useful life, or it can take over from the battery when fast accelerations are necessary or when the battery dies. The flywheel has an additional advantage in that the braking of a car need not be done by frictional brake pads where the energy is lost. As with a battery-propelled car one can use the electrical system to slow or reverse the motor, and the energy of stopping can be returned to the flywheel to be used again. It is estimated this system could increase the range of a conventional battery-operated automobile to over 300 miles. The other alternative is to have the flywheel be the exclusive agent to drive a car.

These applications may seem far fetched but a large company, United Technologies, has been actively investigating the practicality of using flywheels in ways we have indicated. It has constructed a prototype device and has given it to BMW, the German car manufacturer, to try out. There are also smaller companies in the United States that have some designs almost ready to go. They believe that these devices will be on the market early in the 21st century.

Another scenario for using a flywheel may emerge as a result of a small turbine developed by a Los Angeles company, Capstone Turbine. At the moment, this turbine is energized, using gasoline. Yet, it could be run using natural gas or hydrogen, when hydrogen becomes generally available. The turbo generator supplies electrical energy to motors which are at the four wheels of the car. Excess energy is supplied to a flywheel by an electronic controller controlling the blend of flywheel and turbine power for optimum operation. The flywheel could supply surges of energy for rapid acceleration, one of the problems with battery driven cars.

Flywheels can become the energy storage devices if some of the previously described substitutes for fossil fuels become part of the energy picture. For example, if we turn to wind machines, direct solar generators of electricity, or photovoltaic devices, we will need storage facilities because the output of these energy generators is not constant. Batteries have been one of the devices of choice for such storage. But flywheels, if they can be perfected, could also play an important role in such applications.

The materials which are used to make batteries, fuel cells, and flywheels are not necessarily renewable. Their components are not likely to disappear as soon as fossil fuels. Their immediate principal virtue is to improve the environmental burden imposed on us by the extensive use of gasoline-powered trucks and automobiles, thus conserving fossil fuels and improving the air quality at the same time.

Chapter 16

CONSERVATION

After OPEC tripled oil prices in 1973, consumers worldwide reduced their oil consumption. While the increase in the Gross Domestic Product in the United States since then has been close to 40 percent, the energy usage has risen only a small fraction. Much of the decrease of the use of energy has been due to conservation. A smaller part has been due to the United States economy becoming more of a service economy, rather than a manufacturing one.

While conservation was initially driven by purely economic considerations the United States Congress has intervened politically by passing legislation that provides efficiency standards for automobiles and appliances. States, California being the first, have set energy standards for new building construction that are part of the building codes. Legislative action seemed to be called for because in recent years energy use in general and the use of oil in particular has started to increase again. Thus the federal and state governments have now legislated energy conservation as a public policy. This recognition of the importance of conservation has been a partial step toward redressing a past misallocation of resources on the part of the entire society and the government. Traditionally, ten times as much money has been spent on research for increasing the energy supply than has been spent on conservation research. Yet a barrel of oil saved is equivalent to a barrel of oil taken from the ground, and is frequently less costly to come by.

The object of conservation is to have the world's principal stores of energy (oil, coal, and natural gas) last as long as possible. After being forced for economic reasons to reevaluate how energy was being used, the users finally came to realize how profligate their use had been. The lessons, learned as a result of the need to save money, continue to be applied to conserving energy. We shall, in what follows, describe in a general way what has been done and what is being contemplated to carry out the conservation program.

Roughly 35 percent of the energy used in the United States is consumed in residential and commercial buildings. The same percentage is used in manufacturing while 30 percent is used in transportation. In each of these categories the users have found—and continue to find—ways to save energy. We shall not detail all of the ways in which this has been done because that would require a separate book. But by describing some of the results and some likely prospects one gets an inkling of how much energy savings can and will be achieved.

In Chapter 8 we described how flat plate collectors can be used to heat the buildings and provide hot water for the occupants. For existing residential buildings the most significant improvement has not been a high tech discovery. It involves finding and closing gaps in the building envelope to help retain warm air in the winter and cool, air-conditioned air in the summer. The technique involved in finding leaks is to increase the pressure inside of a building being heated and detect the places where air was leaking, either by using smoke sticks or infrared cameras. Some additional low tech procedures, such as adding insulation to the attic or wall, fitting storm windows, wrapping and adjusting the hot water heater, and other similar simple adjustments, provided the greatest savings both in dollars and in energy.

Improvements in the ability of windows to reflect the house's outside heat during the summer and inside heat during the winter can contribute significantly to reducing the energy required for heating and cooling. Double glazed windows, that is, windows consisting of two panes separated by either air, or better still vacuum, provide an easy fix for this problem.

For new construction proper siting of the proposed structure becomes one of the most important energy conservers. To take full advantage of the sun, houses should be aligned roughly from east to west, at an angle of 30 degrees with the east–west line and with many windows facing the south and none facing east or west. One can reduce the overheating in the house so sited in the summer and take advantage of the sun's rays in the winter by placing extended eaves on the south side that would admit the sun's rays to enter the house in the winter when the sun is low in the sky, but would shield it in the summer when it is high.

An emphasis on energy conservation draws attention to the importance of trees and vegetation in decreasing the burden that the seasonal weather places on a house. Another fuel saving device is a thermal mass inside the house. Substantial quantities of concrete, brick, rock, or water are suitable thermal masses because they can absorb the sun's heat during the day, retain it for a long time, and release it at night when the temperature cools. Some home owners have successfully saved energy by building greenhouses on their property and connecting them to the house. These are only few of the many suggestions that have had the effect of conserving energy in homes and residential buildings.

Lighting is one of the largest single expenditure in commercial buildings. For new construction, considerable attention is now being paid to ways of taking advantage of daylight to reduce the energy load supplied by electricity. Large windows properly spaced, skylights, or even "light pipes" have been very effective. Dimming controls for artificial light can also be helpful. Research is being conducted on materials that can change from transparent to opaque in strong sunlight, or when they are heated. Simply changing from incandescent to fluorescent bulbs can effect substantial savings.

Reducing the energy to cool commercial buildings has been achieved by installing electronic sensors and programming computers to regulate the heating and cooling of various parts of a building continuously. Finally, building standards have been formulated that save energy in a variety of other ways. They have become the basis of legislation in several states. In other states many of these standards have been adopted informally by building contractors. Such standards contribute a great deal to the conservation of energy in new construction.

One of the principal energy consumers in the industrial world is motors—devices that convert electrical energy into mechanical energy. In recent years our deeper understanding of the properties of condensed matter has resulted in enormous increases in the efficiency of motors. In addition to reducing the electrical losses caused by currents in the magnets of motors, controls, made possible by the invention of the transistor, have contributed to the increased efficiency of such devices.

Superconductors are materials that can carry a current without dissipating any energy; they offer no resistance to the flow of current. In the wings is the practical development of materials that can become superconducting at a relatively high temperature: the temperature of the boiling point of liquid nitrogen (−195.8°C). This development makes it easier to reach this temperature and maintain it rather than work at the much lower temperature of liquid helium, which is where superconductivity occurred previously. If these materials fulfill their promise there will be a revolution in the improvement of the efficiency of motors used in industry. Substantial energy savings would result over and above what is expended in creating liquid nitrogen and maintaining it in liquid form.

Materials that become superconducting at relatively high temperatures would also hold great promise for reducing the energy required to move trains. Superconductors in the presence of magnets are repelled. If superconductors fulfill their promise their repulsion can be used to levitate trains and eliminate rail travel friction altogether. Such development would bring about a sea change in the dependence of train travel on fossil fuels they now use. Research leading to the development of a levitated train is being carried out in Germany and Japan.

One of Congress' initiatives in energy conservation has been to mandate an average number of miles per gallon for the entire output of every automobile manufacturer. The car manufacturers have succeeded in satisfying the congressional mandate principally by redesigning the engines of their product and using lighter materials in their parts manufacture. As discussed in Chapter 15, there is still room for further developing radically different engines, controls, and car bodies.

Farming is an activity frequently subsumed under the heading of industrial production. But the farmer's work is in some respects different from other industrial activities. The possibility of saving energy in agricultural pursuits deserve some special discussion. Only 7 percent of the population of the United States is engaged in farming whereas more than 90 percent of the population of developing countries such as The Congo are on the farm. The high productivity of the American farmer has been brought about by providing him with a shedful of expensive machinery run on gasoline and diesel fuel and by his saturating the soil with fertilizer. As a result, agriculture consumes about 16 percent of the energy used in the United States, a percentage that includes not only planting, fertilizing, harvesting, but also processing, distribution, and even preparation. What it does not include is the energy used to pump the water that makes arid land suitable for agriculture and the considerable amount of energy required to make the fertilizer, whose primary root stocks are natural gas and petroleum.

In the old days American farmers were much less dependent on fertilizers than they are today. By permitting fields to lie fallow or by planting crops that would renew the fixed nitrogen content of the soil, farmers rarely needed fertilizers. Nowadays, if no pesticides are used, this is called organic farming. It is doubtful that the United States will be able to completely return to this type of farming, for reasons that are political, sociological, and economic. We will have to content ourselves with trying to save as much energy as possible within the framework of modern farming. It is also not clear whether a total return to organic farming could supply us with enough food at reasonable prices to sustain our population.

Fortunately considerable potential exists to save energy without basic changes in present agricultural practices. Many of the savings will not differ much from the savings already discussed in connection with transportation because farm machinery is not so very different from automobiles or trucks. But the really big savings in agriculture will probably come from biological research. Although many genetically engineered farm products have already reached the market, the potential of this technology as an energy saver is yet to be realized. In Chapter 11 we have indicated some of the progress that has been made in creating plants which manufacture their own pesticides and which can fix nitrogen in the soil to be able to dispense with fertilizers. Both of these steps can save enormous quantities of energy if they become widespread. This is the promise of genetic engineering in agricultural energy savings.

Genetic engineering might also help to develop plants that need less water or could grow in brackish or salt water, or in cold weather. Techniques for the improvement of the utilization of the water used in irrigation are being looked into. These researchers should lead not only to the saving of water (which faces as great a potential shortage as do fossil fuels) but also to the saving of energy because the pumping of water is an important sink for energy use in agriculture.

The preservation of food, especially canning, is an important consumer of energy that is a part of food utilization. The ratios of the food energy output of the contents of some canned food to the energy required to can them are: apples in syrup 0.09, carrots in brine, 0.03, baked beans, 0.24, beef 0.34. We hesitate to examine these ratios for preserving pet foods. On the average it takes about ten times as much energy input as one obtains in energy output to can food. If one were to try other methods for preserving food, such as nuclear radiation, which has been proven to be effective as well as harmless, much energy could be saved.

There seems to be no question that the energy policy of the United States is to emphasize conservation. This is made manifest in the encouragement the government is giving to the automobile industry to develop a more efficient motor car, in the efforts being made in implementing the Clear Air Act, where our previous discussion has indicated that the largest expenditures are to make turbines more efficient, and generally urging the citizenry not to waste energy.

The examples we have given of conservation show that the efforts to achieve conservation rest almost exclusively on improving the efficiency of the devices we use which expend energy. As an energy policy the logic of this strategy should be to hope that if efficiency improves we could achieve the same work output with the expenditure of less fuel. Our stores will be conserved and thus will last longer.

It is not clear that even this modest goal can be achieved by depending solely on conservation. In an article in the November/December 1994 issue of *The Sciences*, Herbert Inhaber and Harry Saunders pointed out that except in the short term, conservation efforts often backfire because they lead to increased consumption. This is because increased efficiency is usually accompanied by decreased costs. The authors adduce several examples to support their thesis. Thus James Watt introduced an engine early in the 19th century which was twice as efficient as the Savery engine then in use. This is the sort of step we normally take as a conservation measure. However because the Watt engine made the Industrial Revolution possible, the consumption of coal increased ten-fold in the period between 1830 and 1863.

In 1977 Denmark introduced draconian energy conservation measures. By 1986 Denmark's use of electricity rose by 20 percent despite the tough laws aimed at saving electricity, as a result of the increased use of television and other appliances.

We have already alluded to the 1975 Corporate Average Fuel Economy (CAFE) Act passed by Congress. This act set standards that required an ultimate average fleet energy efficiency for each automobile manufacturer of 27.5 miles per gallon. In 1973 the average was 13.5 miles per gallon. At present it is 22 miles per gallon. Since there are about 200 million cars and trucks in the United States, this act has obviously saved a great deal of gasoline used since it was passed. Despite this attempt at conservation, however, the total amount of automotive fuel has risen since CAFE was passed because of the increased number of trucks of cars and trucks on the road. Gasoline use has increased by about 20 percent.

In all of the cases cited, conservation has had a beneficial effect for society. The increase in the use of fuels came about because the standard of living, the quality of life of the polity improved when an energy source became less expensive. The Industrial Revolution improved the lot of the people; the Danes added television and a host of other appliances for their everyday use; and more cars were used by the Americans with the expectation that the quality of their lives would be improved by these purchases. Improving the standard of living by insisting on conservation is a social benefit of conserving. It does not address what should be the primary reason for conservation, namely to extend the lifetime of nonrenewable fuel sources so necessary for our civilization. To achieve the object of conservation, it is necessary in addition to provide disincentives for using fuel, for example, by increasing its cost by taxation, something our Congress seems loath to do. Its constituency opts for saving money rather than saving fuel.

Our energy policy should aim to replace fossil fuel uses with renewable sources as quickly as possible for the reasons that have been repeatedly given. Real conservation—increasing efficiencies at the same time discouraging usage—has a role to play in such an energy policy. It is important to take advantage of inexpensive sources of energy while they are still readily available. This would give us time to develop the best possible alternative sources. But we defeat the purpose of conservation if we do not use the opportunity to invest and plan well for the future. We do not seem to have the political will to make the financial investment necessary to use this window of opportunity wisely.

Scientists at the Electric Power Research Institute have indicated there is some good reason for encouraging conservation. According to them, the future energy demands for the world will exceed the supplies that could reasonably be expected to be available in the next century. Husbanding the supply is therefore good public policy. In view of these predictions it is surprising the Electric Power Research Institute has not used its substantial political clout to lobby for the development of alternative renewable energy sources.

There are two other reasons our present conservation policy militates against what might be a superior approach. First, by improving the efficiency of equipment designed for fossil fuel, this makes it difficult for alternative sources of energy to compete on a cost basis, and this discourages the use of alternative energy sources. Second, the more spent on our current policy of conservation the less there is made available for encouraging alternative source development.

CHAPTER 17

A HYDROGEN ECONOMY

For some time, people have envisioned an economy where the only source of energy was hydrogen. The idea may have originated in Jules Verne's science fiction novel *Mysterious Island*. There, a shipwrecked engineer says that once they ran out of coal, they would use water as their source of energy. What he meant is the hydrogen in the water could provide the fuel as it did for the bacteria during the early history of Earth.

Hydrogen is a likely fuel. When burned in air, its end products are water and some nitrogen oxides. These oxides, which are potentially pollutants, can be reduced to negligible levels with catalytic heaters. So hydrogen would become the least polluting of all energy sources.

Its versatility makes it a likely candidate for a universal fuel. It has already demonstrated it can be used to fuel space exploration vehicles and to energize the fuel cell that generates their electricity. Hydrogen has also been used, in an experimental way, to power an automobile using a slightly modified conventional engine. Used this way, it is three times as efficient as gasoline. Hydrogen is also a heavyweight with regard to energy density.

Hydrogen's unavailability has prevented it from bestowing its benefits on society. The principal source of hydrogen is in water, where it is tightly bound to oxygen. Until recently, to liberate hydrogen required a great deal of energy, making it a less attractive option. But recent progress in generating electricity using renewable sources that are economical, do not contribute to the pollution of the atmosphere, and do not generate greenhouse gases revived interest in a hydrogen economy. The previous chapters on photovoltaics, direct use of solar energy, wind energy, OTEC, and biomass make the case for paying more attention to hydrogen as a universal fuel. Using power currently generated by fossil fuels to liberate the hydrogen from water would make it too expensive as a fuel. The environmental benefits of the alternative sources plus the benign burden that hydrogen usage places on the environment makes its prospects better if we turn away from fossil fuels.

Given economical sources of available electricity it would require no new technology to develop a hydrogen-producing industry. The electrolysis of water has had a long industrial history. The production of hydrogen from biomass by reforming methanol or ethanol has already been demonstrated.

This fuel has promise as the power source for a Zero Emission Vehicle (ZEV), a development which a number of states are looking forward to. Right now it is contemplated that the ZEV would be battery-powered. Hydrogen would be a superior fuel for two reasons. Its high energy density adds little weight to the car. It is also much easier to refuel. At most it would require replacing one cylinder of compressed gas with another. But an even better prospect is to use the hydrogen in a fuel cell. Fuel cells for vehicles now use methanol as a source of hydrogen. The reforming equipment required to produce the hydrogen makes the installation too heavy for an automobile. But a source of hydrogen that does not require reforming would make possible a car whose energy source is a fuel cell. Such an electric car would not only benefit the environment but would be superior to a gasoline-fueled car.

Many renewable energy sources will probably be used initially for the direct production of electricity, but their potential is somewhat limited by the intermittent character of solar radiation and wind energy, and by the difficulty of using electricity directly for transportation. Using the electricity to make hydrogen creates a mechanism for storing the energy. A steel cylinder of hydrogen is a potential source of energy. The renewable energy generated that is not immediately needed could help decompose water and fill cylinders of hydrogen, which would be used when the sun goes down or the wind stops blowing. This would address one of the shortcomings of some renewable alternative sources of energy.

With all of these significant virtues it would be pleasant to be able to say that the birth of a hydrogen economy is imminent. But such is not the case. There are problems to be solved before we can enjoy its benefits.

Hydrogen storage presents a problem. It can be stored in several ways, none completely safe. With care hydrogen can be at least as safe as many other hazardous substances we use. It can be compressed and stored in tanks. It can be liquified. It can be combined with some metal to form a hydride and then have the hydrogen released by slightly increasing its temperature. Portable tanks tend to be very heavy and awkward to handle but large community tanks can be used as they have been in the past for storing "town gas." Liquefying hydrogen and storing it requires expensive additional sophisticated cryogenic equipment; hydrogen liquifies at the lowest temperature of any element. Making metallic hydrides adds to the expense of using hydrogen.

Hydrogen distribution, if it were to become a much used source of energy, has to be considered. The pipelines currently used to distribute natural gas possibly could be modified to transport the hydrogen. The transmission costs would be about 50 percent higher. There would be some capital expense involved in changing the seals and compressors. Possibly the pipes would need to be replaced with larger diameter ones. (If such a distribution system were in place the estimated cost of distributing hydrogen would be less than the cost of distributing an equivalent amount of electrical energy.) For those places not serviced by pipelines, hydrogen has been demonstrated it can be economically transported in liquified form by ship or rail.

One principal concern about using hydrogen is its safety although in many ways hydrogen is safer than gasoline. Hydrogen is flammable at a 4 percent concentration but gasoline bursts into flames when the concentration is 1 percent. (Natural gas at 5 percent.) The leakage rate of hydrogen is less than that of gasoline. Unlike gasoline, hydrogen is lighter than air and disperses quickly. Thus a flammable or explosive concentration is harder to reach. Nevertheless a harbor full of ships carrying liquified hydrogen looks to many like a catastrophe about to happen. Should something set them off the damage would certainly be enormous. Strictly enforced safety measures would ease this concern.

At present the costs of employing hydrogen as an energy source makes it unlikely we will see a hydrogen economy any time soon. Things will improve as renewable energy sources become cheaper and as the costs of fossil fuels rise. It is possible hydrogen costs will continue to be a problem should such costs be perceived as outweighing the advantages of a cleaner environment and a more secure energy supply.

The 1995 Congress decided in a preliminary way to pay more attention to the possibility of using hydrogen as a fuel. It increased substantially the amount of money devoted to research on hydrogen with an eye to reducing its costs as a fuel and overcoming the difficulties an increase in its usage would present. It seems likely that such trends for hydrogen, and other alternative energy sources, will continue as the problems of fossil fuels worsen and the hard facts of the inevitable exhaustion of such fuels become increasingly manifest.

Chapter 18

ENVOI

Our standard of living, our comfortable way of life has depended on fossil fuels—a seemingly inexhaustible inexpensive source of energy. As a community we have been only vaguely aware of the downside of their use. They are a limited resource and that burning them to extract their energy are defects that our collective unconscious knew but our conscious state thought that no emergency action was required to deal with these problems. Until recently, new stores of fossil fuels continued to be discovered, whose potential yearly output exceeded its potential yearly use. Great progress had been made in lessening the burden that the burning of fossil fuels imposed on the environment. It was expected this problem could ultimately be solved. The question of their eventual scarcity was a problem that would arise so far in the future it could be ignored. This is not an unwarranted reaction for the general public that does not have access to enough relevant data to make a judgment. Prudence would dictate, however, that the leadership of the community should prepare for the eventual scarcity of major sources of energy likely to last only a century at the present rate of use.

Recently a new problem, more urgent than potential scarcity, has appeared arising from the burning of fossil fuels which the vast majority of experts view as one that cannot be solved except by drastically curtailing their use. This is the greenhouse effect. Its environmental consequences were discussed in Chapter 3.

Robert Watson, chairman of the United Nations Panel on Climate Change, in a recent television interview believes that greenhouse warming will have a severe and drastic impact on our way of life in the next century. According to him unusual climatic phenomena such as heavy rains in some areas, heavy snows in others, recent severe droughts and floods are due to greenhouse warming. Burning fossil fuels has increased the height of tides in the islands of the South Pacific, sometimes wiping out houses near the shore. Wildflowers in the Alps have been found at higher elevations. The ice mass in the Antarctic is shrinking. Spring arrives earlier and Fall arrives later. The growing season has lengthened, on the average, by four to 12 days.

Some of the fossil fuel use issues covered in this book were discussed by Hazel O'Leary in the valedictory interview she gave when she retired as Secretary of the Department of Energy in January 1997. She complained the low price of energy in the United States leads to profligate use and entrenches our dependence on fossil fuels. She maintained there is no political will to raise these prices by taxation. Her efforts to nurture alternative fuel and energy technologies such as fuel cells and photovoltaics, during her tenure were hampered by budgetary constraints. She found the prevailing view in the country was that somehow, miraculously, the market for these technologies would materialize when needed. She recommended that we institute a program to support these technologies, possibly subsidized by the government.

Much remains to be investigated. Thus there is an urgency in formulating an energy policy aimed at addressing the problem of ultimate scarcity and greenhouse warming. It is doubtful any one technology could replace fossil fuels, so that research would have to be conducted on many fronts. Alternative sources alone may not be able to generate enough power for our needs. This question has not been adequately investigated because the alternatives were not economical while fossil fuels were an affordable energy source.

As noted earlier, the 1997 International Conference in Kyoto, Japan tried to address one of these problems. The best agreement the participating countries could reach leaves one with the uneasy feeling that there is no constituency that favors long term safety over short term profit.

To make substantial progress in dealing with the energy problem requires the public to demand a change in our energy policies. President Clinton has indicated that he intends to use his presidency to seek such public support. It has been our hope that this book will help mobilize public opinion. People have to become more sensitive to the dangers future generations face. They should not be frightened by the propaganda that change will lead to economic disaster. There may be a difficult future for those whose fortunes depend on fossil fuels use in our economy, but even these will have many years to adjust. No one expects that fossil fuel use will disappear overnight. Once the door is opened for research into new technologies, economic opportunities will abound. One encouraging indication of progress happened as this book was being written. The National Energy Laboratory announced it had invented a semiconductor device that converts light into hydrogen suitable for fuel in one step. Light goes in and hydrogen comes out. The first device by the laboratory had an efficiency twice any previous devices that could convert light energy to hydrogen. When Brazil turned to substitutes for gasoline it created hundreds of thousands of new jobs and reduced its trade deficits.

Those opposed to encouraging the development of renewable sources of energy by government subsidies argue that global warming is not a real effect but a computer artifact. They believe the ultimate shortage of fossil fuels is a problem so far into the future that conservation alone can solve it. They are trying to convince the public that the experts are Cassandras. Even if this were true it is obvious that a prudent policy must allow for these experts to be right. The development of a prudent policy seems for many reasons not to be taking place.

INDEX

Illustrations are denoted by page references in **bold type**.